Impact of Oxidative Stress on Diabetes Mellitus and Inflammatory Bowel Diseases

Impact of Oxidative Stress on Diabetes Mellitus and Inflammatory Bowel Diseases

Jana Varvařøvská, Josef Sýkora,
Zdenìk Rušavý, Jaroslav Racek,
Silva Lacigová, Konrad Siala
and Rudolf Štìtina

Nova Science Publishers, Inc.
New York

For permission to use material from this book please contact us:
Telephone 631-231-7269; Fax 631-231-8175
Web Site: http://www.novapublishers.com

Library of Congress Cataloging-in-Publication Data

Impact of oxidative stress on diabetes mellitus and inflammatory bowel diseases / Jana Varvaøovská ... [et al.], authors.
p. ; cm.
Includes bibliographical references and index.
ISBN 978-1-60456-152-4 (softcover)
1. Diabetes--Pathophysiology. 2. Inflammatory bowel diseases--Pathophysiology. 3. Oxidative stress--Pathophysiology. I. Varvařøvská, Jana.
[DNLM: 1. Diabetes Mellitus--physiopathology. 2. Oxidative Stress--physiology. 3. DNA Damage--physiology. 4. DNA Repair--physiology. 5. Inflammatory Bowel Diseases--physiopathology. WK 810 I344 2007]
RC660.I445 2007
616.4'62--dc22

2007047826

Published by Nova Science Publishers, Inc. ✢ New York

Contents

Preface

Formation of reactive oxygen species (ROS) is a natural process during oxidative metabolism. ROS play an important role not only in pathological processes of human organism as usually presented but less attention is paid to their proper important role in cell signaling, biosynthesis or non-specific antiinfectious defence. Overproduction of ROS during numerous pathological situations in presence of insufficient antioxidant protection leads to substantial oxidative changes of lipids, proteins, sugars, and also DNA. Protection against ROS is assured by different extracellular or intracellular antioxidant mechanisms as studied during last decades. Antioxidant enzymes rectifying the oxidative damage are studied with regard to their different activities and usefulness in body protection. Their genetic polymorphisms are certainly involved in different response to oxidative stress.

Special attention should be devoted to the topic of oxidative nuclear and mitochondrial DNA damage and its restoring via DNA repair process, especially base excision repair (BER). A large scale of antioxidant enzymes is involved in correction of DNA oxidative damage. Natural trend of worsened DNA repair is usually associated with aging. Other pathologies related with deficient DNA repair are susceptibility to carcinogenesis (lack of apoptosis control) or degenerative diseases.

Oxidative stress is involved in the pathophysiology of diabetes mellitus (DM – oxidative stress of mainly metabolic origin) and inflammatory bowel diseases (IBD – oxidative stress of mainly inflammatory origin). In spite of confirmed OS in DM or IBD, the substantial information about the intensity of DNA repair and its possible relationship to the disease course and development of chronic complications is missing.

Pilot studies completed both in adult and pediatric patients with DM or IBD confirmed an increased oxidative stress as well as oxidative DNA damage examined with comet assay. The surprising findings were ascertained in intensity of DNA repair (analysed with modified comet assay). DNA repair process was stimulated in Type 1 diabetes adults without diabetic microvascular complications and still more in diabetic children with short disease duration. On the other hand, adults with Type 2 diabetes had substantially increased oxidative DNA damage and extremely low DNA repair. This finding could be the link to increased susceptibility of Type 2 DM patients to cancerogenesis. Patients with IBD (Crohn´s disease - both children and adults) had similar tendencies in OS intensity and oxidative DNA damage and repair but less intensive.

Large population studies in DM or IBD studying intensity of OS and expression of DNA repair enzymes are needed in order to get the correlation between individual repair enzymes expression and the long-term course and occurence of complications in DM or IBD.

Keywords: oxidative stress, reactive oxygen species, antioxidant defense, oxidative DNA damage, DNA repair, Type 1 diabetes mellitus, diabetic microvascular complications, Crohn´s disease

Chapter I

Introduction

1.1. ROS under Physiological Conditions

Formation of reactive oxygen species (superoxide, hydrogen peroxide, hydroxyl radical, ROS) is a natural process during oxidative metabolism. ROS play an important role not only in pathological processes of human organism as usually presented but less attention is paid to their proper important role in cell signalling, biosynthesis or non-specific antiinfectious defence.

Triplet-state molecular oxygen undergoes process of univalent reduction and formation of superoxide anion ($O_2^{-\cdot}$) nonenzymatically by redox-reactive compounds in the mitochondrial electron transport chain and enzymatically via enzymes such as NAD(P)H oxidases and xanthine oxidase. Further conversion of labile superoxide is into hydrogen peroxide (H_20_2) by superoxide dismutase. In the presence of reduced transition metals (e.g. ferrous or cuprous ions), hydrogen peroxide can be converted (via Fenton reaction) into the highly reactive hydroxyl radical ($^{\cdot}OH$). Alternative conversion of hydrogen peroxide into water is mediated by the enzymes catalase or glutathione peroxidase. In the glutathione peroxidase reaction, glutathione is oxidized to glutathione disulfide, which can be converted back to glutathione by glutathione reductase in an NAD(P)H-consuming process [1](figure 1). Superoxide and related ROS formed via NAD(P)H oxidase have important physiological functions such as signal transduction from various membrane receptors, control of ventilation, smooth muscle relaxation, and enhancement of immunological functions, control of erythropoietin production and other hypoxia-inducible functions, cell apoptosis.

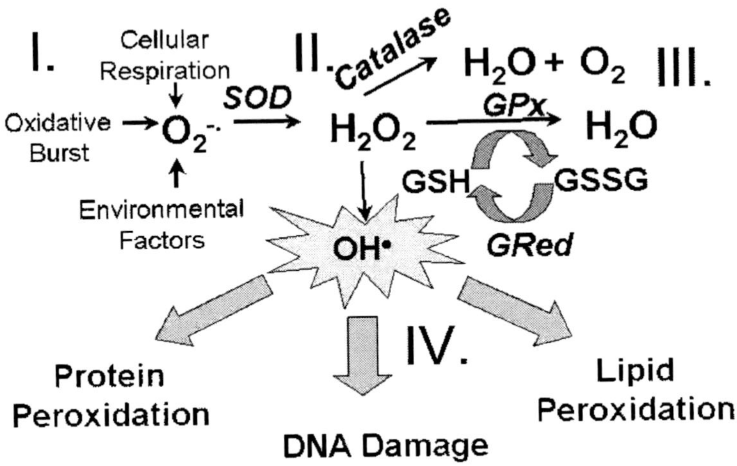

by G.P.Eckert
www.biozentrum.uni-frankfurt.de/Pharmakologie/index.html

I. Generation of superoxide, II. Formation of hydrogen peroxide
III. Detoxification of hydrogen peroxide by antioxidant enzymes
IV. Protein, lipid and DNA damage by ROS (hydroxyl radical) (by G.P. Eckert)

Figure 1. Scheme of oxidative stress in human cell.

Under physiological conditions, ROS formation and concentration in biological cells and tissues is determined by the balance between their rates of production and their rates of clearance by various antioxidant compounds and enzymes. Main nonenzymatic compounds are such as α-tocopherol, ascorbate, β-carotene, glutathione, amino acids, albumin, and the most important enzymes involved in antioxidant defence are superoxide dismutase SOD, glutathione peroxidase GPx, glutathione reductase, glutathione transferase, and catalase [1].

Oxidant-antioxidant balance is very important for cells and tissues life and metabolism. A short-term increase in ROS concentrations or a decrease in the activity of one or more antioxidant systems stimulate cell signal cascades and lead to stimulated gene expression and higher production of proteins or antioxidant enzymes (oxidative stress response) [2]. The original balance between ROS production and ROS scavenging capacity is restored and a so-called redox homeostasis is re-established.

ROS play important role in receptor-mediated signalling pathways. Growth factors, cytokines or other ligands trigger ROS production in non-phagocytic cells through NADPH oxidase stimulation. Such ROS production mediates further signal transduction including stimulation of protein tyrosine kinases, mitogen activated protein kinases (MAPK) cascade, oxidative activation of protein kinase C as well as activation of transcription factor AP-1 important for differentiation processes and also transcription factor NF-κB, implicated in inflammatory reactions, growth control and apoptosis.

ROS are involved in the defence against environmental pathogens and immunological response. In phagocytic cells (macrophages, neutrophils), NADPH oxidase produces ROS after stimulation with microbial products, lipoproteins or by cytokines such as interferon-γ, interleukin-1β, or interleukin-8. The stimulated massive production of ROS is called oxidative burst and plays an important role as a first line of defence against environmental pathogens. The following product created by combination activities of NADPH oxidase and myeloperoxidase is hypochlorous acid (HClO), a potent antimicrobial and oxidant agent. ROS production in lymphocytes is mediated by the enzyme 5-lipoxygenase (5-LO) and may influence their response to cytokines. Exposure of T lymphocytes to relevant concentrations of environmental ROS amplify signalling cascades even after mild stimulation and in case of massive ROS production substantially enhances the lymphocyte reactivity and stimulation.

ROS produced during oxidative phosphorylation in mitochondria serve as sensor in carotid body for changes in oxygen concentration and also for expression of erythropoietin. ROS also induce the adherence of leukocytes to endothelial cells in post-capillary venules. ROS are crucial for normal functions of cardiac and vascular cells including vascular tone.

Though the physiological and favourable effect of ROS on a normally functioning cell, there is a propensity for ROS to evade antioxidant defences and to modify cell components – lipids, proteins, sugars, and finally DNA. If ROS impact is short, there is no serious damage to the cell. On the other hand, prolonged ROS effect leads to cell apoptosis, cellular dysfunction, or neoplastic shift.

1.2. ROS Under Pathological Conditions

If ROS overproduction during numerous pathological situations in human medicine remains stable and long-lasting, it could be taken as *basic pathological mechanism for many different diseases and complex processes* as aging, neoplasias, infectious diseases, autoimmune disorders, ischaemia-reperfusion syndromes (myocardial infarct, brain stroke, injuries, organ transplantation), atherosclerosis, diabetes mellitus, toxic factors of different origin, irradiation, extreme physical activity and *systemic and organ specific diseases* as pancreatitis, glomerulonephritis or interstitial nephritis, inflammatory bowel diseases, hepatitis, different dermatological diseases (atopic dermatitis, psoriasis, infections, pulmonary affections (asthma, dust diseases, ARDS), rheumatoid arthritis, demyelinating diseases [3].

When considering the origin of non-regulated ROS production, it is necessary to refer to *the most important situations:*

a) Mitochondrial electron transport chain (ETC), b) inflammatory processes with excessive stimulation of NDAPH oxidases and lipoxygenases, c) ischaemia and reperfusion with xanthine oxidase stimulation, d) endogenous or exogenous toxic factors including radiation.

Oxidative stress (OS) is then defined as the imbalance between high ROS production and insufficient antioxidant (AO) defence. Once formed, ROS deplete antioxidant defences, rendering the affected cells and tissues more susceptible to oxidative damage. Lipids, proteins and DNA are the cellular targets for oxidation, leading to changes in cellular structure and function. The order of preference for modification depends upon a number of factors, such as location of ROS production, relative ability for the biomolecule to be oxidised, and availability of metal ions. Oxidative changes of lipids and proteins may be removed via normal turnover of the molecules; damage to DNA is required to be repaired. Prolonged oxidative stress leads to serious injuries of cell structures and may be finished by cell apoptosis, dysfunction or uncontrolled cell division and neoplasia. *Diabetes mellitus and inflammatory bowel diseases represent 2 model diseases with important oxidative stress in their pathogenesis and they will be discussed later.*

1.3. Antioxidant Defence of Human Organism

There are several approaches how to distinguish components of defence against oxidative stress. The schedule underlines the process of ROS formation or degradation and hence the defence against them [3, 4].

a) Substances preventing the formation of ROS

are mainly the compounds binding reduced transition metals (e.g. ferrous or cuprous ions) such as acute-phase proteins (transferrin, ferritin, lactoferrin, haptoglobin, hemopexin, ceruloplasmin), or chelating agents, also inhibitors of enzymes catalysing the formation of ROS (allopurinol blocking xanthine oxidase).

b) Substances removing previously formed ROS or restoring reducing agents are either intracellular and extracellular enzymes such as superoxide dismutase, catalase, glutathione peroxidase, glutathione reductase and transferase, or non-enzymatic endogenous and exogenous substances such as glutathione, α-tocopherol, ascorbate, β-carotene, amino acids, uric acid, bilirubin, albumin, and polyphenols.

c) Repair systems removing molecules damaged by ROS are principally enzymes degrading modified proteins, or oxidized lipids (phospholipases, glutathione peroxidases), and finally the complex of enzymes repairing DNA damaged by ROS.

1.4. Some Remarks to Antioxidant Enzymes and Substances from the Recent Literature

Antioxidant Enzymes are Directly Implicated in Human Protection against Oxidative Stress

Their properties, their known polymorphisms and their corresponding phenotype are target of many studies.

Superoxide dismutase SOD transforms superoxide into hydrogen peroxide. Three isoforms (cytoplasmic CuZn SOD or SOD1, mitochondrial Mn SOD or SOD2, tetrameric CuZn extracellular EC-SOD or SOD3) exist and have been studied for many years *in vitro* and *in vivo*. Relationships among different SOD isoforms and human diseases were found. Down regulated or mutant SOD1 has been associated with stroke, ischaemia/reperfusion syndrome, neurodegenerative disorders (Parkinson's disease) and familial amyotrophic lateral sclerosis with 58 known mutations in the CuZn SOD gene (chromosome 21q22.1). Loss or reduction of the SOD2 activity has been associated with neurodegeneration, heart failure and also death during neonatal period. SOD3 has been detected in several compartments such as the plasma, lymph, ascites, cerebrospinal fluid and lung and is related to inflammation, hyperoxic pulmonary disorders, and focal cerebral ischaemia [5]. The present efforts are done to introduce into clinical practice SOD mimetic drugs [6], either selective (macrocyclics) or non-selective such as manganese metalloporphyrins, manganese salen complexes [7, 8]. Another possibility seems to be dietary delivery system using wheat gliadin biopolymers carrying plant SOD extracts [9].

Glutathione peroxidases GPx scavenge peroxides generated in cells. There are four isoforms with different tissue localisation. Cytosolic GPx1 is ubiquitously expressed in human organism, particularly in erythrocytes, kidney, and liver. Cytosolic GPx2 is expressed primarily in gastrointestinal tissues and plays an important role in protection against ingested lipid peroxides. GPx3 is an extracellular enzyme reducing hydroperoxides of more complex lipids. It is mainly present in plasma, but also in lung, heart, kidney and placenta. GPx4 is detected both in the cytosol and associated with membranes. This enzyme reduces peroxidised phospholipids and cholesterol in membranes [5] (figure 2).

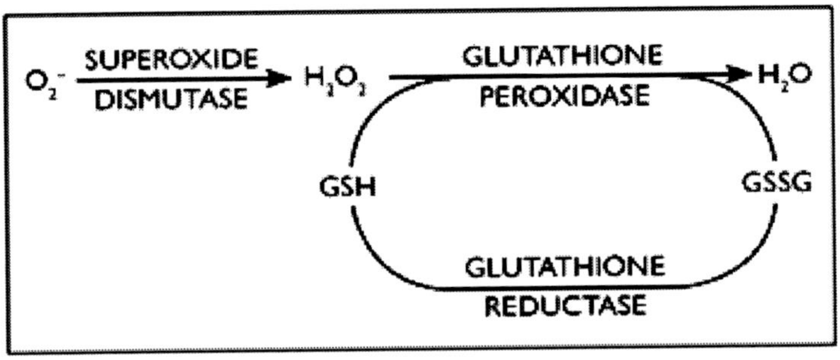

Figure 2. Glutathione peroxidase and glutathione reductase action in antioxidant defence and glutathione metabolism.

Glutathione reductase GSR has two isoenzymes (cytosolic and mitochondrial) and its function is to restore glutathione into reduced form [5].

Catalase decomposes hydrogen peroxide into water and oxygen. This enzyme is particularly abundant in erythrocytes, hepatocytes and the kidney. There are several mutations/polymorphisms in the catalase gene (chromosome 11p13) associated with acatalasemia in erythrocytes, but phenotypes associated with acatalasemia are not prominent [5].

Glutathione transferases GST exist in six cytosolic isoforms and one microsomal GST, the latter now designated membrane associated proteins involved in eicosanoid and glutathione metabolism (MAPEG). GSTs play an important role in detoxification of xenobiotics (chemical carcinogens, environmental pollutants, antitumor agents) and in inactivation of endogenous hydroperoxides and unsaturated aldehydes. GSTs catalyse the conjugation of 4-hydroxynonenal, the degradation product of lipid peroxidation, with glutathione

[5, 10, 11]. GST expression is regulated by the cellular redox status and represents a sensor able to transmit the redox variation to the apoptosis machinery. Cytosolic GST isoenzymes are broadly cytoprotective, but their genetic polymorphisms can increase susceptibility to carcinogenesis [12, 13] and inflammatory disease such as chronic obstructive pulmonary disease, asthma, pancreatitis, autoimmune hepatitis, rheumatoid arthritis [5, 14, 15, 16, 17, 18].

It is obvious that cells exposed to oxidative stress express stress-induced genes or genes encoding antioxidant defences [19], and their polymorphisms are certainly involved in various responses to oxidative stress [5].

1.5. Non-Enzymatic Substances of Endogenous or Exogenous Origin

Non-Enzymatic Substances of Endogenous or Exogenous Origin are of great importance for antioxidant defence. Some of them are newly synthesized and classified as antioxidant enzymes mimetics and were mentioned concurrently with antioxidant enzymes. Further new antioxidants are analogues of naturally existing substances.

Glutathione (GSH - γ-glutamylcysteinylglycine) draw attention in context with different diseases. GSH is the major intracellular non-protein antioxidant and is present in all mammalian tissues, where it plays a crucial role in the detoxification of free radicals. Its synthesis serves to maintain a reduced cellular environment. The liver is the main site of inter-organ glutathione synthesis and supplies GSH into circulation and systemic distribution, but the GSH production occurs also in other body cells such as erythrocytes, GIT epithelium, and bronchiolar epithelium. GSH pool is drawn on for three major applications: for direct free radical scavenging and as an antioxidant enzyme cofactor (glutathione peroxidases GPx, glutathione transhydrogenases reconverting dehydroascorbate to ascorbate), and as cofactor for the glutathione transferases GST. Total glutathione has two forms – the free form which is mainly in its reduced form (GSH) and the form bound to proteins (gluthathionylated proteins) [11, 20, 21]. Conversion of GSH reduced form into oxidised glutathione (GSSG) occurs during oxidative stress and can be reverted to the reduced form by the action of glutathione reductase GSR mentioned above (figure 3). This enzyme uses as its source of electrons the coenzyme NADPH coming mainly from the pentose phosphate shunt. Low GSH, high GSSG, and a higher GSSG/GSH ratio indicates the presence of oxidative stress and consumption of GSH [11].

Glutathione has also profound importance for cellular homeostasis and for diverse cellular functions such as protein synthesis, enzyme catalysis, transmembrane transport, receptor action, intermediary metabolism, cell maturation, and apoptosis [11, 20].

The importance of glutathione and its disturbances has been highlighted in numerous articles [11, 22, 23]. GSH deficiency certainly plays the role in liver diseases such as cirrhosis, hepatitis and non-alcoholic steatohepatitis [20, 21]. Depletion of glutathione was also found in severe acute pancreatitis [16]. Alterations in alveolar and lung GSH metabolism is widely recognized as a central feature of many inflammatory lung diseases such as idiopathic pulmonary fibrosis, acute respiratory distress syndrome, chronic obstructive pulmonary disease (COPD), and asthma [24, 25, 26]. In patients with cystic fibrosis (CF), CFTR channel is no more taken for a primarily „chloride channel“ but recognized as a channel that also controls the efflux of other physiologically important anions, such as glutathione and bicarbonate. CF mutations in fact cause a primary dysfunction in the reduced GSH system, e.g. CF mutations significantly decrease GSH efflux from cells and this leads to deficiency of GSH in the epithelial lining fluid of the lung, as well as in other compartments, including immune system cells and the gastrointestinal tract. Following this, other antioxidants are depleted and this may play a role in initiating the over-inflammation characteristic of cystic fibrosis [26, 27]. The role of glutathione is also alleged in neurodegenerative diseases (Parkinson's disease, Alzheimer's disease), in atherosclerosis, and in diabetes whose topic will be discussed later [20, 22].

Strategies for cellular glutathione repletion include mainly dietary interventions. Natural dietary glutathione undergoes hydrolysis by intestinal and hepatic gamma-glutamyltransferase [28], and so food sources or supplements that increase glutathione must either provide the precursors of glutathione, or enhance its production by some other means. Natural foods that boost glutathione levels are fresh and frozen fruits and vegetables (asparagus, broccoli, avocado, spinach, walnuts), fish, and meat [29]. N-acetyl cysteine takes on significance as a dietary GSH source. N-acetylcysteine has been used in several indications, such as the treatment of paracetamol (acetaminophen) intoxication [30], prevention of contrast-induced nephropathy after intravascular angiography [31], treatment of chronic obstructive pulmonary disease [32]. Its application in the treatment of severe sepsis and septic shock remains controversial [33, 34, 35]. N-acetylcysteine amide is currently studied as another possible substance for treatment of neurodegeneration and other oxidation-mediated disorders [36].

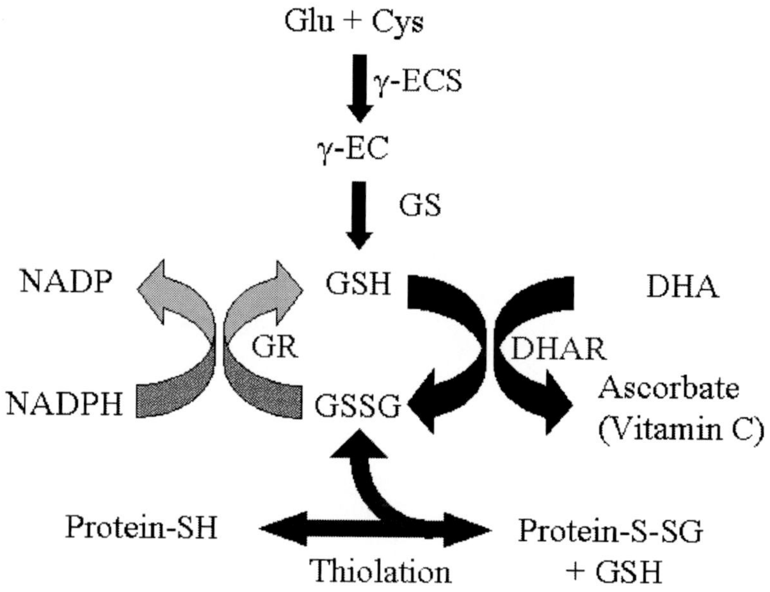

GR-glutathione reductase (NADPH-dependent), GS-glutathione synthase, γ-ECS- gamma-glutamylcysteine synthase,

DHA-dehydroascorbate, DHAR-dehydroascorbate reductase, GSSG- oxidized glutathione, GSH-reduced glutathione

Figure 3. Metabolism of glutathione in the cell.

Thioredoxin (Trx)

Thioredoxin (Trx) is a small multifunctional protein involved in cellular thiol-redox processes. It contains two redox-active cysteine residues in its active sites. The protein is stress-inducible and enhanced by various types of stresses including ROS. The thioredoxin is kept in the reduced state by the flavoenzyme thioredoxin reductase, in a NADPH-dependent reaction. It acts as an electron donor to peroxidases and ribonucleotide reductase and the administration of recombinant Trx protein, induction of endogenous Trx, or gene therapy can be a therapeutic modality for oxidative stress-associated disorders [37,38, 39, 40].

Coenzyme Q10 (CoQ)

Coenzyme Q10 (CoQ) known as ubiquinone, CoQ, and vitamin Q10 is a fat-soluble vitamin-like quinone present in every cell of the body. It is naturally present in different amounts in foods (mostly high in beef meat, soy oil, sardines, mackerel, and peanuts), but it is also synthesized from its precursors Q1-9 or *de novo* synthesized from acetyl coenzyme A in liver. CoQ-10 is found throughout the body in cell membranes, especially in the mitochondrial membranes and is particularly abundant in the heart, lungs, liver, kidneys, spleen, pancreas and adrenal glands.Q10 is a mobile electron carrier from NADH dehydrogenase complex to cytochrome b-c_1 complex within the mitochondrial electron transport chain enabling the H^+ flow out of the mitochondrial matrix. Thus electrochemical proton gradient across the inner membrane is formed and finally drives the H^+ flow through the ATP synthase, that uses the energy of the H^+ flow to synthesize ATP as a carrier of free energy [39]. Q10 is also a powerful antioxidant both on its own and in combination with vitamin E. CoQ is present not only in mitochondria, where bound to mitochondrial membranes serves also as an antioxidant. It is also present in cell membranes where in its reduced form ubiquinol can neutralize a lipid peroxyl radical. The CoQ radical ubisemiquinone is stable and its formation is a normal intermediate between ubiquinone (CoQ) and ubiquinol ($CoQH_2$), and is restored to a non-free-radical state by the respiratory chain Q cycle. Ubisemiquinone and ubiquinol can also regenerate the vitamin E tocopheroxyl radical by electron donation and this is very important as vitamin E neutralizes lipid peroxyl radicals far more readily and is less hydrophobic, allowing it to more freely move throughout the mitochondrial membrane. CoQ is more confined to the centre of the phospholipids layer. Its importance as an endogenously produced lipid-phase antioxidant is indicated with its high concentration not only in cell membranes, but also in Golgi vesicles and microsomes [4, 41, 42]. Supplementation of CoQ is recommended in patients with Parkinson's disease, mitochondrial encephalomyopathies, with chronic heart failure [41], as CoQ is needed for energy production and antioxidant protection. The problem in question is the supplementation of CoQ in patients treated with statins as HMG-CoA reductase is also the first committed rate-limiting step for the synthesis of a range of other compounds including steroid hormones and ubiquinone. Currently, there is insufficient evidence from human studies to link statin therapy unequivocally to pathologically significantly decreased tissue CoQ levels [43]. Cosupplementation with vitamin E and CoQ reduces circulating markers of inflammation as proved on animal models and this example can be used for further studies dealing with vascular diseases [44].

Vitamin C

Vitamin C is essential for human organism. Under physiological conditions, vitamin C predominantly exists in its reduced form, ascorbic acid (AA), it also exists in trace quantities in the oxidized form – dehydroascorbic acid (DHA). AA functions as a cofactor for enzymes involved in the biosynthesis of collagen, carnitine, norepinephrine, and in the amidation of hormones. In plasma and cells, AA is a powerful antioxidant, quenching ROS. During the process of quenching free radicals, ascorbate donates an electron, becoming the unstable intermediated ascorbyl radical that can be reversibly reduced back to ascorbate via glutathione oxidation. Ascorbyl radical can also donate a second electron and be converted to DHA that can efflux from the cells via the glucose transporters and functions as a readily transportable form of vitamin C. As mentioned above, ROS activate transcription factors, such as NF-κB, that are important in host defence, inflammation, and apoptosis. AA directly quenches ROS intermediates involved in the activation of NF-κB, and is oxidized to DHA, which directly inhibits IκBα kinase (IKKβ) and thus NF-κB cannot be activated [45].

Vitamin E

Vitamin E is a fat-soluble vitamin present in many foods, especially certain fats and oils. There are α, β, γ, and δ forms, among them α-tocopherol (α-TOH) is the most abundant and active form with the referring term vitamin E. It is a natural, lipophilic radical-scavenging antioxidant and is described as the most important lipophilic antioxidant in vivo, which has been the subject of many extensive studies [46]. Vitamin E localization in the lipophilic domain of the membranes and lipoproteins inhibits lipid peroxidation, i.e. α-TOH reacts with peroxyl radicals much faster than lipids do and thus it may scavenge the majority of peroxyl radicals before they attack lipids. The efficacy of α-TOH is better when a radical scavenging occurs near the membrane surface. If the radical went deeper into the interior of the membranes, scavenging activity of α-TOH decreased significantly [47]. α-TOH is converted to α-TO· radical which may undergo several reactions. Preferentially, it is reduced by ascorbate in order to regenerate α-TOH provided that α-TO· radical is localized in the membranes and LDL. If the radical goes deeper into the interior of cell or is localized in the LDL core, ascorbate is unable to reduce the radicals. The α-TO· radical can be reduced by other reductants, such as ubiquinol and polyphenolic compounds. Vitamin E may exert its effect by acting as a radical-scavenging antioxidant whose activity depends on many factors (localization, mobility, presence of other collaborating antioxidants). These factors influencing vitamin E efficiency and impact on oxidative stress can explain certain controversies in the antioxidant studies with vitamin E. Nevertheless, vitamin E remains a perspective antioxidant in combination with synergistic antioxidant compounds [46, 47].

Phytochemicals

Phytochemicals are a wide range of different naturally plant compounds, such as carotenoids, polyphenols including flavonoids. Their antioxidant effects depend on their chemical structure and lipophilic or hydrophilic character. Carotenoids are plant pigments and animals do not synthesize them. They are responsible for the red, yellows, and orange colour of fruits and vegetables, their number in nature exceeds 600 and can be grouped into carotenes, xanthophylls and lycopene. Carotenoids have been intensively studied in the last decade and as proved by several epidemiological studies, they showed certain preventive or inhibitory effect on atherosclerosis, degenerative diseases, cancer and other diseases [48, 49, 50]. The antioxidant activity of carotenoids arises as a result of the ability to delocalise any unpaired electrons. β-carotene has an excellent ability to quench singlet oxygen without degradation and reacts with free radicals such as peroxyl radical, hydroxyl radical. This is the way that assures lipid protection from peroxidative damage. Polyphenols, among them flavonoids existing for example in green tea, black currant, blackberries, form stable phenolic radicals with ROS, regenerate vitamin C and E and bind transition metals [4]. Resveratrol is a naturally occurring polyphenol (phytoalexin) produced by some higher plants in response to injury of fungal infection. Resveratrol is found in grapevines, also in peanuts and mulberries. It has several activities which include inhibition of the oxidation of low-density lipoproteins, blockade of superoxide anion and hydrogen peroxide in macrophages stimulated by lipopolysacharides (LPS), decrease of arachidonic acid release induced by LPS or by exposure to superoxide or hydrogen peroxide, glutathione-sparing activity thanks to hydroxyl-radical scavenging activity, suppression of lipid peroxidation both by chelation of copper and by scavenging of the free radicals. Resveratrol action is explained by its influence on NADPH-dependent oxidases, xanthine oxidase, 5-lipoxygenase and 15-lipoxygenase, myeloperoxidase. It seems to be a good candidate protecting the vascular walls from oxidation, inflammation, platelet aggregation, and thrombus formation, i.e. atherosclerotic process [51]. The French paradox (low incidence of coronary heart in France) is attributed to the complex antiatherosclerotic role of grape polyphenols regularly consumed in red wine. Isorhapontigenin (ISO), a new resveratrol analogue, has been isolated from Belamcanda chinensis (blackberry lily). Its chemical structure is very similar to that of resveratrol. ISO significantly inhibits malondialdehyde formation induced by hydrogen peroxide and the increase in lipid peroxidation, ISO also inhibits oxidative DNA damage [48, 52].

Figure 4. shows the cellular distribution of some enzymatic and non-enzymatic antioxidants.

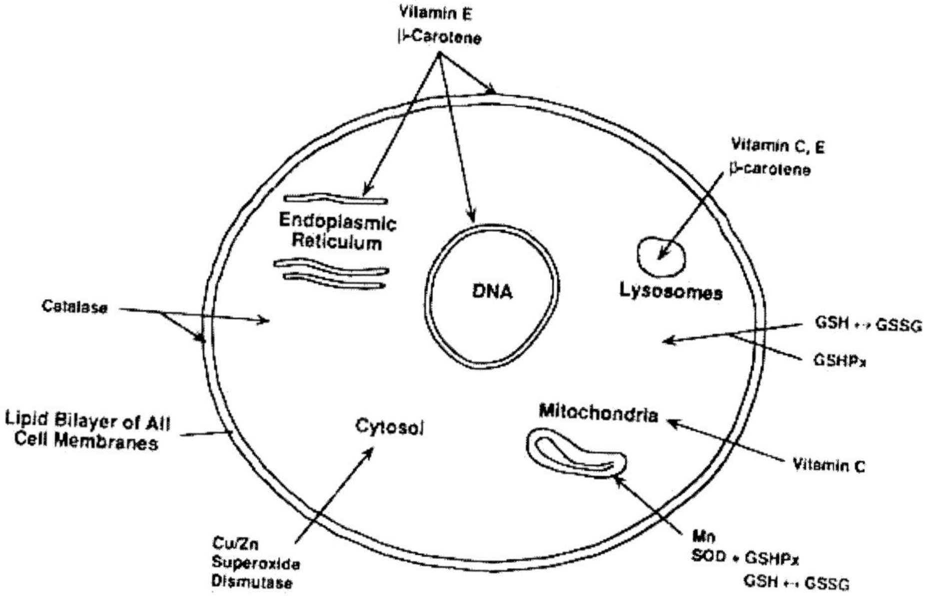

Available from: www.medscape.com, by RH Demling and L De Santi

Figure 4. Cell structure and location of different endogenous and exogenous antioxidants.

1.6. Catalytic Antioxidants – New Therapeutic Possibilities

As mentioned above, superoxide $O_2^{\cdot-}$ and hydrogen peroxide H_2O_2 participate in the formation of several reactive species that have been directly implicated in tissue damage. Cellular antioxidant enzymes as SODs and catalase detoxify ROS. Unfortunately, their large size, short circulating half-life, and antigenicity are not suitable for therapeutic use in free radical diseases. To overcome many of these limitations, an increasing number of low molecular-weight catalytic antioxidants have been developed. They are designed with redox-active metal centres that catalyse the dismutation reaction of $O_2^{\cdot-}$ and/or H_2O_2 by a mechanism that is similar to the mode of action of the active-site metals of SOD and catalase. An ideal mimetic is stable and non-toxic and its size and charge of a mimetic can be exploited to target crucial cellular sites of oxidative stress. Three classes of metal-containing catalytic antioxidants, i.e. macrocyclic (M40403), salen (EUK series), and metalloporphyrin mimetics (AEOL series) seem to be important for future clinical use as shown on *in vitro* and *in vivo* models. Their potential includes lung fibrosis, chronic obstructive lung diseases, asthma, acute respiratory distress syndrome, bronchopulmonary dysplasia, shock, myocardial infarction, liver failure, colitis, diabetes, ischaemia-reperfusion, transplantation, arthritis, amyotrophic lateral sclerosis, neurodegenerative diseases, stroke [7, 8].

Chapter II

DNA Oxidative Damage and DNA Repair

2.1. Nuclear DNA Damage and Repair

DNA is constantly exposed to damaging agents from both endogenous and exogenous sources. If these lesions are not repaired, they can lead to mutations and result in cellular dysfunction with either further uncontrolled cell proliferation, or apoptosis. In order to maintain the integrity of the genome, a complicated network of DNA repair pathways remove the majority of deleterious lesions. There are six major repair pathways of DNA repair [53, 54, 55]. The number of human DNA repair genes exceeds 140 and keeps growing [56, 57]. Direct reversal (DR) consists of repair proteins acting directly on a damaged base and re-establishing the correct structure in situ without removing the damaged nucleotide. Exogenous agents such as ionising radiation, tobacco smoke, cytostatics generate larger DNA lesions with abnormal structure or abnormal chemistry (24-29 nucleotides) which are repaired via nucleotide excision repair (NER). NER steps involves a) damage recognition, b) binding of a multi-protein complex at the damaged site, c) double incision of the damaged strand several nucleotides away from the damaged site, on both the 5′and 3′sides, d) removal of the damage-containing oligonucleotide from between the two nicks, e) filling in of the resulting gap by a DNA polymerase and f) final ligation. The genes encoding many of the human NER proteins were first identified in genetic complementation studies of the human DNA repair disease – xeroderma pigmentosum (XP) with mutations in any of 7 genes. In addition to XP,

Cockayne´s syndrome (CS) and trichothiodystrophy (TTD) are two other human genetic disease with defects in NER [58, 59].

The mismatch repair (MMR) is a post-replicative DNA repair system contributing to the fidelity of replication and the maintenance of the genome. Recombinational repair (RER) restores deleterious double-strand breaks via homologous recombinational repair (HRR) or non-homologous end-joining (NHEJ). Genes and proteins related to HRR are implicated in chromosomal stability and are critical in maintaining cellular resistance to ionising radiation [60]. Ataxia teleangiectasia (AT) and Nijmegen breakage syndrome (NBS) are chromosome instability syndrome with mutated genes, such as ataxia teleangiectasia mutated (ATM - chromosome 11q23) or NBS respectively. AT is an autosomal recessive disease and affected patients develop cerebellar ataxia, teleangiectasia, combined cellular and antibody immunodeficiency, sinopulmonary infections, and malignancies. The Nijmegen breakage syndrome (NBS – chromosome 8q21) is also an autosomal recessive disease characterized by microcephaly, growth retardation, immunodeficiency and cancer predisposition [61, 62]. Germline and somatic mutations of certain HRR genes contribute inevitably to carcinogenesis (e.g. BRCA1, BRCA2 proteins)[63].

Finally, base excision repair (BER) is the most important pathway for oxidative DNA damage. Free radicals, most notably ·OH, react with DNA compounds and form radicals of DNA bases and the sugar moiety. Further reactions of these radicals result in a variety of final products. Major products of oxidative damage to the DNA bases are numerous, among them thymine glycol, 5-formyluracil, 5-hydroxyuracil, 5-hydroxycytosine, 8-hydroxyadenine, 8-hydroxyguanine, 2,6-diamino-4-hyroxy-5-formamidopyrimidine. In addition to sugars modified by oxidative stress (e.g. 2-deoxyribonolactone), unaltered sugar moieties within DNA, called base-free sites (or apurinic/apyrimidinic site – AP site) are formed from the elimination of modified DNA bases due to the weakening of the glycosidic bond. These lesions can result in single strand breaks.

As ROS are continuously generated in all organisms, the oxidized bases, single-strand breaks and AP sites in DNA are repaired primarily via the DNA base excision repair pathway (BER) (figures 5, 6). NER is implicated as a back-up system for BER, so oxidative DNA lesions containing oligonucleotide are removed via NER and is taken as a back-up system for BER.

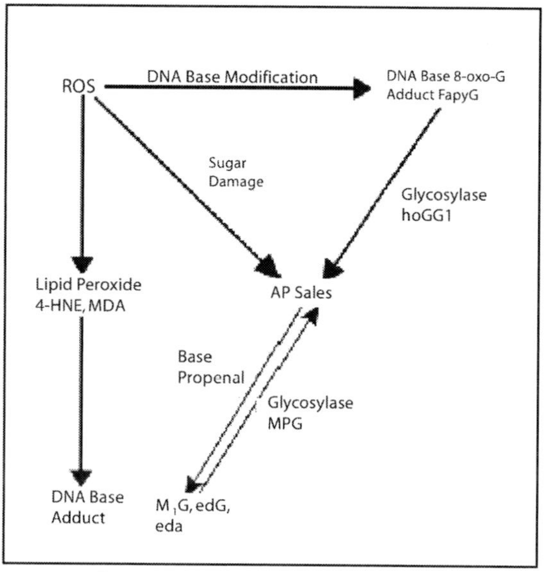

ROS-reactive oxygen species, 4-HNE- 4-hydroxy-2-nonenal, MDA-malondialdehyde, M1G pyrimidopurinone, edG ethenodeoxyguanosine, edA etheno-deoxyadenosine, hOgg1 human 8-oxoguanine DNA glycosylase 1, MPG N-methylpurine DNA glycosylase, FapyG 2,6-diamino-4-hydroxy-5-formamidopyrimidine, 8-oxo-G 8-oxo-7,8-dihydro-2′-guanosine

Figure 5. Schematic diagram of types of DNA damage induced by ROS (published by Powell, CL, Swenberg, JA and Rusyn, I. in Cancer Letters, 2005, 229, 1-11).

The developing knowledge about DNA repair processes is bringing the detailed information about purine- and pyrimidine-derived lesions and their removal from DNA. 8-OH-guanine glycosylase OGG1 has 2 major forms hOGG1-1a being targeted to the nucleus and hOGG1-2a being targeted to the mitochondrial inner membrane and protecting mitochondrial DNA. OGG1 is a constitutively expressed gene and its expression can vary with tissue source and between individuals. MutY homologue (hMYH) glycosylase repairs mismatches 8-OH-Gua:A and 8-OH-Gua:G. Human MutT homologue (hMTH1) degrades 8-OH-dGTP to 8-OH-dGMP and pyrophosphate. N-methylpurine-DNA glycosylase (MPG protein or 3-methyladenine-DNA glycosylase) has a relatively broad range of substrates including 8-OH-Gua and 3-methyladenine by BER mechanism. Endonuclease III homologue (hNTH1) is involved in the repair of oxidized pyrimidines. Another guanine-derived 2,6-diamino-4-hydroxy-5-formamidopyrimidine (FapyGua) is removed via hOGG1 or NEIL1 (glycosylase excising FapyGua). Adenine-derived lesions such as 8-hydroxyadenine (8-OH-

Ade) are removed also via hOGG1 when 8-OH-Ade pairs opposite cytosine. 2-OH-adenine is the substrate for hMTH1. 4,6-diamino-5-formamidopyrimidine (FapyAde) is released by the activity of hNTH1 and also NEIL1 glycosylase. The same enzyme is targeted at 5-hydroxycytosine (5-OH-Cyt), 5-OH-uracil lesions and thymine glycol. Thymine glycol could be removed also by thymine glycol glycosylases (primarily mitochondrial TGG1 and nuclear TGG2) [2, 64].

Removal of a damaged nucleotide by a glycosylase forms an AP site. Repair of AP sites inhibits both the mutational and cytotoxic effects of nucleic acid loss. AP endonucleases (APEX1, APEX 2) recognize abasic site and cleave the phosphodiester DNA backbone. Subsequent step is up to the DNA polymerase β which removes the deoxyribose phosphate and adds a nucleotide. Finally, DNA ligase III in partnership with the XRCC1 protein (ligase accessory factor) seals the nick. Cells deficient in polymerase β showed hypersensitivity to monofunctional alkylating agents. The described pathway is called short patch-BER (figure 6) [54, 59].

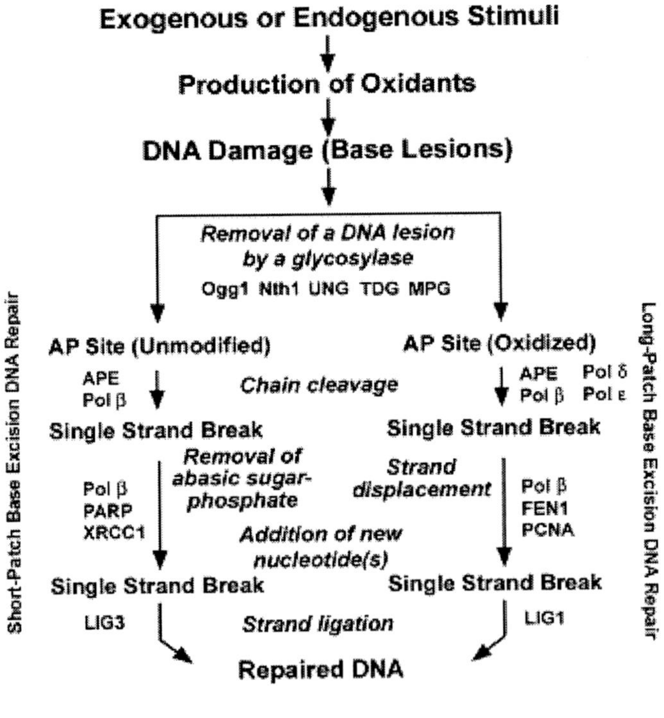

Figure 6. Base excision repair pathway (published by Powell, CL, Swenberg, JA and Rusyn, I. in Cancer Letters, 2005, 229, 1-11).

The second BER pathway - long patch-BER is a little bit complicated and involves the activities of glycosylase, AP-endonuclease I, replication factor C, proliferating cell nuclear antigen (PCNA), DNA polymerase δ or ε, flap-endonuclease 1 (FEN1 – 5′nuclease) and DNA ligase I. This involves removal and re-synthesis of a 2-6 nucleotide patch (figure 6)[54, 59, 64].

Whichever pathway is used, the product secreted from the cell will be the modified base, because they both pathways involve a glycosylase activity. These excreted oxidized bases could be measured as a marker of endogenous oxidative DNA damage i.e. oxidative stress [2]. There is a large scale of methods for measuring oxidative DNA damage: chromatographic assessment of oxidized purines and pyrimidines, comet assay (or single cell gel electrophoresis SCGE) for determination of single-strand breaks [65], when lesion-specific endonucleases are included, allows detection of oxidized purines and pyrimidines. Slot-blot assays enables the measurement of abasic (AP) sites and with addition of lesion-specific endonucleases allows detection of oxidized purines and pyrimidines [66]. For example, an intensively studied 8-OH-guanine was measured in blood and especially in urine as a marker of oxidative stress in different diseases. Parkinson's and Alzheimer's disease, multiple sclerosis, amyotrophic lateral sclerosis, breast cancer, colorectal cancer, haematological disorders, chronic hepatitis, haemochromatosis, Wilson's disease, cystic fibrosis, lung and gastric cancer, rheumatoid arthritis, systemic lupus erythematosus, Down's syndrome, diabetes mellitus [2].

Repair enzymes are constitutively expressed, their expression and/or activity may be modulated in response to various potential genotoxic stress or in concert with the cell cycle. Levels of oxidative bases in DNA are the consequence of a balance between lesion induction from radical processes and repair. Reduced repair will result in elevated lesion and an increased risk of disease. Genetic polymorphisms in repair proteins could also affect the relative propensity of individuals to develop various pathologies including even human inherited disease. Biallelic mutations in the BER DNA glycosylase MYH were found in an autosomal recessive syndrome of adenomatous colorectal polyposis with very high risk of colorectal cancer [67]. Several studies focusing the tissue expression of BER enzymes, and their relationship to diseases have been accomplished or are in motion using animal models or human samples [66, 68, 69]. Variable population distribution of polymorphisms in DNA repair genes and inter-individual differences in the DNA-repair efficiency are referred to the different occurrence of oxidative stress related diseases in the world [69, 70].

2.2. Mitochondrial DNA Damage and Repair

BER Repair Pathway Comes about not Only in Nucleus, but Also in Mitochondria

Mitochondria (originated from maternal egg) are present in many copies in mammalian cells, except for mature erythrocytes (figure 7). Each mitochondrion has a few identical copies of its own circular DNA, which codes for 13 mitochondrial enzymes involved in cellular respiration (using a slightly different genetic code than does nuclear DNA).

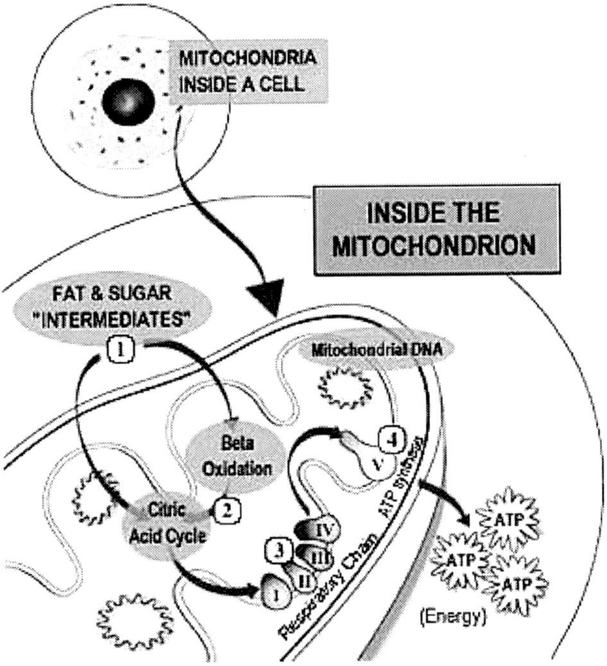

available from http://www.mitoresearch.org/images/q64bigmito.jpg

Figure 7. Metabolic pathways in mitochondrion.

Nuclear genes code for several other mitochondrial proteins. In somatic cells, mitochondrial DNA mutates about 10 to 20 times faster than nuclear DNA, because mitochondria are a major source of reactive oxygen species. These mutations accumulate with age in postmitotic tissues such as neurons, and in cardiac and skeletal muscle. As studied in multiple reports [71] and brilliantly summarized in one of the latest reports [72], there is an evidence that multiple DNA repair pathways exist also in mitochondrion (DR, BER, MMR, RER). BER pathway in mitochondrion shares similar steps like in nuclear BER but taking into account that the responsible enzymes have nuclear and mitochondrial isoforms (OGG1, TGG, APE, MYH). The presence of the only strand could also contribute to the observed increased mutation frequencies of mtDNA compared to nDNA. Though this fact, other non-mutated copies of mtDNA are present and thus mitochondrial function does not necessarily be compromised. Repair of mtDNA and nDNA in principle could be very similar with the difference being that nDNA repair pathways involve a larger number of proteins.

If oxidative DNA damage is not repaired completely, several situations can come about. Many oxidative base lesions are mutagenic, other can induce conformational changes in DNA. Lesions as thymidine glycol can block replication one nucleotide before and after the lesion. The length of repetitive sequences of DNA (microsatellites) is constant in normal cells, but microsatellite instability (i.e. variable number of repeats) derives from oxidative DNA damage. If the DNA oxidative lesions are present in the transcribed regions of genes they can lead to mutation. But if they are localized in promoter, elements they can affect transcription factor binding [4, 64].

2.3. Poly(ADP-Ribose) Polymerase

Poly(ADP-Ribose) Polymerase 1 (PARP) Plays an Important Role in Oxidative DNA Damage Repair

Poly(ADP-ribose) polymerase 1 (PARP-1) is the 113-kDa enzyme, mostly found in nucleus, member of the PARP enzyme family consisting of PARP-1 and several recently identified poly(ADP-ribosylating) enzymes. PARP-1 represents the best studied PARP enzyme and is responsible for more than 90% of the cellular poly(ADP-ribosyl) capacity. The process of poly(ADP-ribosyl)ation plays the role in a variety of physiological and pathophysiological phenomena, such as signalling of DNA damage induced by ionising radiation, alkylating agents and oxidants, DNA-base excision repair, regulation of genomic stability in cells under genotoxic stress, transcriptional regulation, stimulation of nuclear proteasomal function, ageing and longevity, further in pathophysiology of acute and chronic inflammatory disorders (e.g. Crohn´s disease), diabetes mellitus, ischaemia-reperfusion damage in tissue, septic and haemorrhagic shock [73]. The primary signal that cause PARP-1 activation is the recognition of single or double strand DNA breaks caused by various genotoxic agents inclusive oxidative stress in the course of different pathological situation as described above. Activated PARP-1 dimerizes at recessed DNA ends and cleaves NAD+ into nicotinamide and ADP-ribose which forms long branching (ADP-ribose)$_n$ polymers. The latter product is then attached to glutamate or aspartate residues of nuclear acceptor proteins, including PARP itself (PARP automodification) (figure 8).

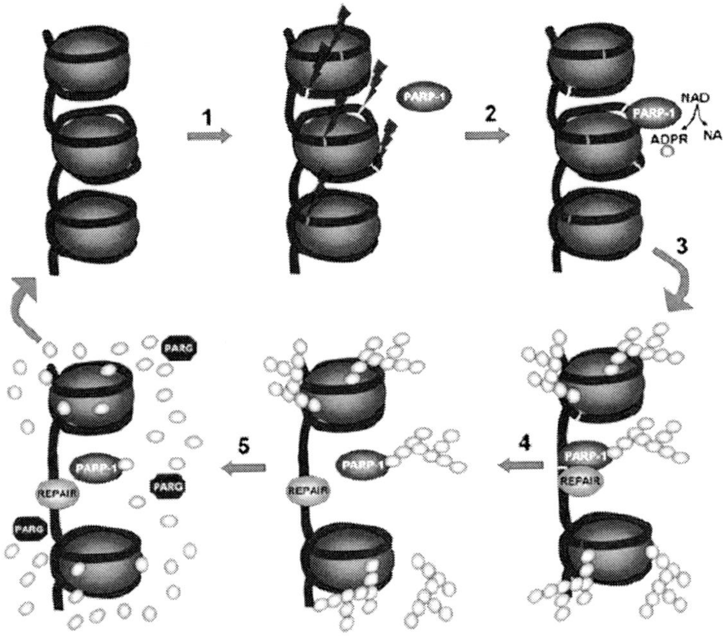

Free radicals and other agents as UV light, ionising radiation or alkylating agents may cause DNA strand breakage. PARP-1 binds to recessed DNA ends resulting in the activation of the enzyme. Active PARP-1 cleaves NAD+ into ADP-ribose (ADPR) and nicotinamide (NA). Then PARP-1 polymerises ADPR covalently linked to itself and histones leading to recruitment of various DNA repair adapter and effector proteins. Poly(ADP-ribosyl)ation of histones causes relaxation of chromatin structure rendering the site of DNA damage accessible to the repair machinery. Automodified PARP-1 detaches from DNA and the repair machinery seals broken DNA ends. PARG (poly(ADP-ribose) glycohydrolase) decomposes the polymer and the chromatin regains its original structure.

Figure 8. DNA damage-induced activation of PARP-1 (published by Virag L. Current Vascular Pharmacology, 2005, 3, 209-214, with the kind permission of the author).

As described by Virag [74], by conferring negative charges to acceptors, this covalent protein modification alters the physico-chemical properties of the modified target proteins and thus regulates their function. The set of polymer acceptor proteins includes various DNA repair proteins, transcription factors, histones, other proteins, and PARP-1 itself. This PARP-1 automodification represents a negative feedback mechanism causing the down regulation of PARP-1 enzyme activity, the poly(ADP-ribosyl)ation is reversible and the enzyme poly(ADP-ribose)glycohydrolase (PARG) degrades the polymer (figure 8)

As described, nuclear ADP-ribosylation induced by the presence of DNA strand breaks stimulates DNA repair, and finally leads to the cell recovery from moderate DNA damage (figure 8). If DNA damage is excessive and unrepairable, PARP is overactivated and mediates a futile energy-consuming cycle inactivating the stress response of the cells [73, 74].PARP-1 depletes the cellular stores of its substrate NAD+ , and in turn NAD+ resynthesis also consumes ATP. Hence, NAD+ and ATP depletion as two key energy metabolites leads to cell death, apoptosis or necrosis (figure 9). The switching from apoptosis to necrosis by PARP-1 overactivation may aggravate tissue damage, as leakage of cellular content from necrotic cells is pro-inflammatory. It is necessary to stress that the intensity of cellular poly(ADP-ribosyl)ation decreases with age either in humans or in animals [73].

Except for the role of PARP-1 in regulation of genomic stability and cell survival, it has also an impact on gene expression. Gene interaction with PARP-1 had either positive or negative effects on the specific transactivation process. An important example is the involvement of PARP-1 in the regulation of NFκB, the specific transcription factor involved in the control of numerous genes, including those of many cellular genes involved in immune responses and inflammation. PARP- has also been implicated in the regulation of activator protein (AP-1)-driven transcriptional activity (figure 9). The activites of PARP shows certainly the role in many pathological processes, such as acute ischaemia-reperfusion damage, allergen-induced asthma, septic or haemorrhagic shock, and chronic including cancer, cardiac disorders, Parkinson´s diesease, inflammatory bowel disease and diabetes mellitus. It is necessary to emphasize the fact that the function of PARP-1 and poly(ADP-ribosyl)ation as a regulator of inflammatory signal transduction and transactivation shows marked cell-type and stimulus dependency [75]. This finding might certainly play the role in possible usage of envisaged PARP inhibitor therapy as will be mentioned later.

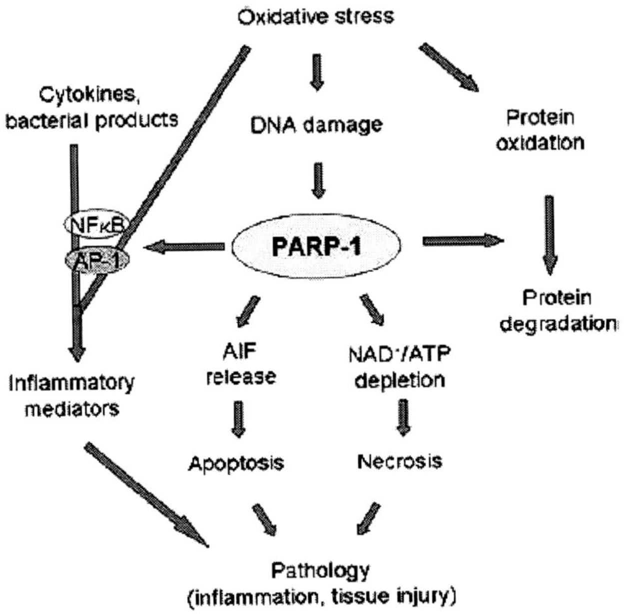

The central role of PARP-1 in oxidative stress-related pathology. Oxidative-stress-induced DNA strand breakage triggers the activation of PARP-1, leading to apoptosis inducing factor (AIF) release from the mitochondria and AIF-mediated, caspase-independent apoptotic cell death NAD/ATP consumption and consequent necrotic cell death. Oxidative stress also stimulates activation of redox-sensitive transcription factors such as NfκB and activator protein 1 AP-1, key regulators of inflammatory cytokines and chemokines.

Figure 9. The role of PARP-1 in oxidative stress-related pathology (published by Virag L. Current Vascular Pharmacology, 2005, 3, 209-214, with the kind permission of the author).

A close association was described between PARP-1 and p53 protein in DNA damage-induced apoptosis [74, 76]. p53 protein is a nuclear protein that plays an essential role in the regulation of cell cycle, specifically in the transition from G0 to G1, the cell cycle arrest at the G1/S transition (which prevents the manifestation of unrepaired chomosome alterations), and also on cell cycle arrest at the G2/M transition, which inhibits the distribution of defective genomes. P53 also initiate apoptosis after the introduction of irrepairable DNA damage. At this point, both PARP-1 activation and p53 activation occur rapidly following DNA damage. The activation of these two mediators is likely to be linked to each other, as PARP-1 has been shown to interact with and poly(ADP-ribosyl)ate p53 [74, 76].

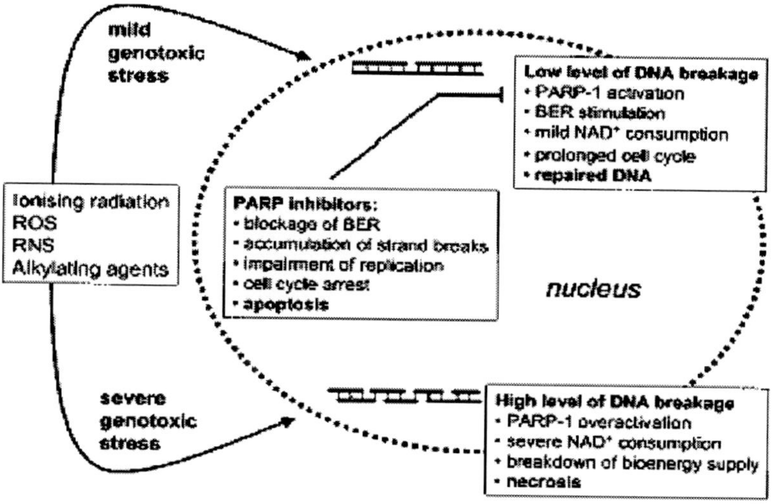

Oxidative-stress-induced DNA strand breakage triggers the activation of PARP-1. If the DNA breakage is low, BER is stimulated, cell cycle is prolonged (in cooperation with p53 protein activation) and DNA is finally repaired. If DNA breakage is high, PARP-1 is over activated, severe NAD+ consumption is in progress and apoptosis or even necrosis is started. PARP inhibitors may block the inflammatory pathways but also BER.

Figure 10. DNA breakage and the activity of PARP inhibitors (published by Beneke, S; Diefenbach, J; Bűrkle, A. Poly(ADP-ribosyl)ation inhibitors: promising drug candidates for a wide variety of pathophysiologic conditions. Int J Cancer, 2004, 111, 813-818, with the kind permission of the author A. Bűrkle) .

The described activities of PARP-1 in DNA base excision repair, in apoptosis or cell necrosis induction, stimulation of inflammatory mediators via NFκB and AP-1 transcription factors, and protein degradation have been taken into consideration in several studies with PARP inhibitors on animal models or cell cultures [77](figure 10) and recently also in clinical trials on oncological or diabetic patients [78, 79, 80]. Most PARP inhibitors developed to date have structural similarity to the substrate NAD+, thus competing for the active centre of the enzyme. The compounds are non cytotoxic on their own at concentrations necessary to achieve PARP inhibition. PARP inhibitors are used for cancer treatment with regard to the effect of PARP inhibition and potentiation of cell necrosis [77]. Another indication is to turn down the stimulation of pro-inflammatory mediators like in chronic obstructive lung disease, diabetes mellitus, or congestive heart failure, and Parkinson´s disease [80]. The question remains what side effects might be observed after a long PARP inhibitor application with regard to the blockage of DNA repair in non-cancerous diseases.

2.4. Diseases Related with Oxidative Stress and DNA Damage

Natural trend of worsened DNA repair is usually associated with aging. Aging as a constant and inevitable process has been related to oxidative stress for many years. Telomere shortening related to a finite number of replication (Hayflick´s model) is observed in normal human somatic cells and is considered as one of the mechanisms of the senescence (in this case – replicative senescence) [81].

Aged cells show age-dependent accumulation of 8-oxoguanine in the nucleus and age-dependent decrease in the activity of OGG1 imported into the nuclei and into mitochondria [82, 83]. Also AP endonuclease showed age-associated decreased activity, i.e. less intensive BER corresponds with the theory of aging associated with oxidative stress [83, 84].

Another cause of aging is related to mutation of human WRN gene. The role of normal Werner protein (or RECQ3 helicase) is to interact with many proteins involved in BER, and in this way to facilitate BER by unwinding BER intermediates. Mutated WRN gene give rise to a rare autosomal recessive genetic disorder Werner syndrome with premature aging, predisposition to cancer and early onset of aging symptoms such as osteoporosis, ocular cataracts, greying, loss of hair, diabetes mellitus, atherosclerosis [85, 86, 87].

Facts related to aging and polymorphisms of antioxidant enzymes have not yet been fully established. Reports on mutations of SOD causing amyotrophic lateral sclerosis are known [5].

The mitochondrial theory of aging states that senescent cellular changes are related to the balance between inherited healthy mitochondrial DNA and repaired mtDNA and the load of age-related unrepaired mutations in this DNA [84]. Eventually, damage or loss of mitochondrial DNA prevents postmitotic cells from regenerating new mitochondria, thereby reducing the production of ATP and leading to cell death. Oxidative damage of mtDNA, accumulating with age, should affect mitochondrially encoded proteins, but the high number of mitochondrial genomes allows a certain degree of heteroplasmy (different genomes within a mitochondrion) without effects on phenotype. Clonal accumulations of damaged/mutated mtDNA within individual cells up to homoplasmy of mutant mtDNA, which could be either neutral with regard to phenotype or which could cause substantial phenotype alterations. The degree of age-induced mutations in mitochondrial DNA varies in people and may be under genetic control. Some mitochondrial DNA haplotypes are more common in centenarians than in the general population [88, 89, 90]. Mitochondrial oxidative stress should be considered a hallmark of cellular aging. Mitochondrial ROS production mediates signal transduction pathways sensitive to oxidative stress and this would be involved in the onset and development of age-related degenerative diseases and that limit the mean life span. The second pathway is related to oxidative damage to mitochondrial DNA, proteins, and lipids and would determine the maximum life span [91].

Other pathologies related with oxidative stress and DNA damage are susceptibility to carcinogenesis (lack of apoptosis control), and other non-cancerous pathological conditions, such as degenerative diseases (Alzheimer's disease, Parkinson's disease), metabolic diseases (haemochromatosis, Wilson disease, diabetes mellitus), inflammatory diseases of infectious and non-infectious origin (atherosclerosis, autoimmune, and allergic diseases), ischaemia-reperfusion injury. The intensity and quality of DNA repair are studied particularly in tissue sections, or tissue cultures, human leucocytes, or on animal models in some pathological conditions, such as malignant tumours [2, 64, 92, 93, 94], neurodegenerative diseases [68, 95], atherosclerosis [88, 96, 97], inflammatory bowel diseases [98], diabetes [99, 100], and in workers exposed to genotoxic effects of different chemicals [101].

The foregoing condensed summary dealing with oxidative stress and oxidative DNA damage and particularly BER serves for further elucidation of specific changes in two models of free radical diseases – diabetes mellitus and inflammatory bowel diseases.

Chapter III

Oxidative Stress and Pathophysiology of Diabetes Mellitus and Inflammatory Bowel Diseases

Among human diseases mentioned above, diabetes mellitus (DM) and inflammatory bowel diseases (IBD) represent 2 chronic lifelong diseases that could be present as 2 model free radical diseases. ROS in DM are produced via different pathways, among them the most important is the ROS overproduction by the mitochondrial ETC and glucose autoxidation, i.e. *oxidative stress of predominantly metabolic origin with consequent inflammatory processes*. In IBD, mainly Crohn's disease, excessive stimulation of otherwise tightly regulated NADPH oxidases during inflammatory process contributes to the ROS formation and the presence of *oxidative stress of predominantly inflammatory origin.*

3.1. Diabetes Mellitus

is defined as a group of chronic, metabolic heterogeneous disease marked by high levels of sugar in the blood. It results from insufficient action of insulin in consequence of its absolute or relative deficiency or both. Diabetes as a life-long disease is accompanied with complex disturbances of sugar, lipid and protein metabolism. The clinical features of diabetes are well-known, such as polyuria and polydipsia, nycturia, loss of weight, fatigue, loss of consciousness even coma, acetone in breath, recurrent infection of skin and urogenital system, later occurs the development of microvascular complications (retinopathy even blindness, nephropathy even renal failure, peripheral and autonomic neuropathy) and atherosclerosis with accompanied hypertension, heart attack, brain stroke, claudication [102](figure 11).

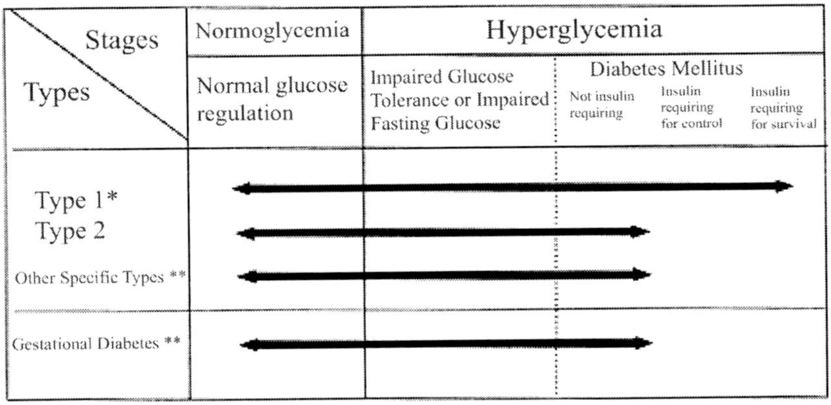

Figure 11. The Expert Committee on the Diagnosis and Classification of Diabetes Mellitus ADA (Diabetes Care 26: S5-20, 2003).

Type 1 diabetes (T1DM) develops as a result of autoimmune insulitis in genetically predisposed individuals, typically is diagnosed in childhood, adolescence, or early adulthood. Autoimmune destruction of the ß-cells of the pancreas is reflected with markers of the immune destruction of the ß-cell include islet cell autoantibodies (ICAs), autoantibodies to insulin (IAAs), autoantibodies to glutamic acid decarboxylase (GAD_{65}), and autoantibodies to the tyrosine phosphatases IA-2 and IA-2ß. Also, the disease has strong HLA associations, with linkage to the DQA and B genes, and it is influenced by the DRB genes, these genes being either predisposing or protective. These patients are also prone to other autoimmune disorders such as Graves' disease, Hashimoto's thyroiditis, Addison's disease, vitiligo, and pernicious anaemia. Incidence of the disease is variable with regions, and in Europe an increasing tendency from the South to the North is reported with the highest number of 50/100000 children up to 15 years/year in Finland. The age-related incidence has its peak at the age of 13-15 years. T1DM dominates in childhood and adolescence. The disease onset in adulthood is no more rare (LADA - latent autoimmune diabetes of adults) and has been intensively studied. As T1DM is insulin dependent, different insulin intensified therapies are applied. Besides insulin, also diet, and regular physical activity are necessary for T1DM successful treatment.

Type 2 diabetes (T2DM) is characterized by peripheral insulin resistance with an insulin-secretory defect that varies in severity. For type 2 diabetes to develop, both defects must exist: all overweight individuals have insulin resistance, but only those with an inability to increase beta-cell production of insulin develop diabetes. 90% of patients who develop type 2 diabetes are obese. Patients with type 2 diabetes often do not need treatment with oral antidiabetic medication or insulin if they lose weight. There are probably many different causes of this form of diabetes, and it is likely that the proportion of patients in this category will decrease in the future as identification of specific pathogenic processes and genetic defects permits better differentiation among them and a more definitive subclassification. Autoimmune destruction of ß-cells does not occur, and patients do not have any of the other specific causes of diabetes. Both genetic and environmental factors (above mentioned obesity) contribute to insulin resistance. Adipose tissue plays a central role in the pathogenesis of insulin resistance that induces a compensatory increase in β cell mass, which results in normal glucose levels. In other people, β cell response is not adequate and leads to impaired glucose tolerance or even type 2 diabetes. The presence of chronic hyperglycaemia i.e. glucotoxicity induces multiple defects in β cells, including early decreases in glucose-stimulated insulin secretion, and late irreversible changes in insulin-gene transcription and β cell mass with consequent low insulin production. The world epidemic of obesity and thus the related increased incidence of T2DM threatens developed countries and represents a serious problem to public health. Worldwide prevalence of T2DM is estimated as 150 million cases in 2002 (Report of a WHO Meeting).

Other specific types of diabetes includes genetic defects of the ß-cell (monogenetic defects in ß-cell function - MODY), point mutations in mitochondrial DNA, genetic defects in insulin action (mutations of the insulin receptor), diseases of the exocrine pancreas (acquired processes include pancreatitis, trauma, infection, pancreatectomy, cystic fibrosis and haemochromatosis), endocrinopathies (e.g. acromegaly, Cushing's syndrome, glucagonoma, pheochromocytoma), drug- or chemical-induced diabetes, infections (rubella, coxsackievirus B, cytomegalovirus, adenovirus, and mumps), uncommon forms of immune-mediated diabetes (systemic lupus erythematosus and other autoimmune diseases), other genetic syndromes sometimes associated with diabetes (Down's syndrome, Klinefelter's syndrome, and Turner's syndrome. Wolfram's syndrome).

Gestational diabetes mellitus (GDM) is defined as any degree of glucose intolerance with onset or first recognition during pregnancy.

The increasing number of diabetic patients all over the world is an important problem of public health. Quality of life of diabetic patients, chronic diabetic complications, treatment of all types of diabetes and its complications at vast expenses represent important challenges not only for medical science, but also for human society.

3.1.1. ROS and Diabetes

Hyperglycaemia is common for all types of diabetes. This metabolic finding is the key for all important consequences related to oxidative stress. High constant influx of glucose under hyperglycaemia via GLUT 1 transporters leads to ROS overproduction at the mitochondrial level (figure 12). In diabetic cells with high glucose inside, there is more glucose being oxidized in the citrate cycle (TCA cycle), which in effect pushes more electron donors (NADH and $FADH_2$) into the electron transport chain. As a result of this, the voltage gradient across the mitochondrial membrane increases until a critical threshold is reached. At this point, electron transfer inside complex III is blocked, causing the electrons to back up to coenzyme Q, which donates the electrons one at a time to molecular oxygen, thereby generating superoxide (figure 13). Overproduction of superoxide is the first and key event in the activation of all other pathways of hyperglycemic damage involved in the pathogenesis of diabetic complications [103, 104, 105, 106, 107], such as formation of advanced glycation end-products AGEs, stimulation of protein kinase C and transcription factor NFκB, polyol pathway and hexosamine pathway (figures 12, 14, 15, 16, 17). The cited work of Brownlee presents an integrated approach to the oxidative stress in diabetes.

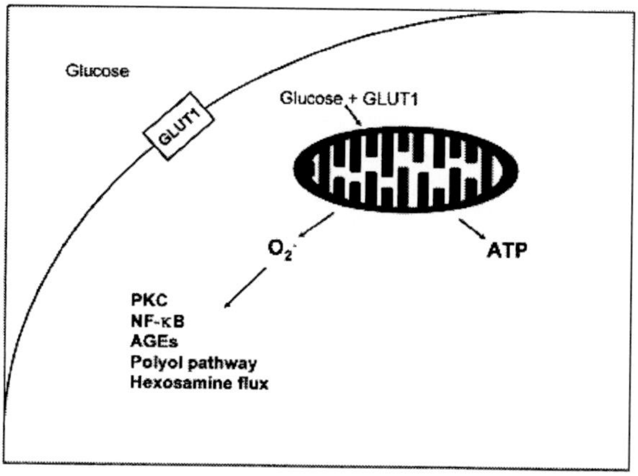

In endothelial cells, glucose can pass freely through the cell membrane in an insulin-independent manner via GLUT1. Intracellular hyperglycaemia induces overproduction of superoxide at the mitochondrial level. Overproduction of superoxide is the first and key event in the activation of all other pathways involved in the pathogenesis of diabetic complications, such as polyol pathway flux, increased advanced glycosylation end product (AGE) formation, activation of protein kinase C (PKC) and NF-κB, and increased hexosamine pathway flux.

Figure 12. The role of increased flux of glucose into the cell and mainly to the mitochondrion (published by Ceriello, A. New insights on Oxidative Stress and Diabetic Complications May Lead to a "Causal" Antioxidant Therapy. Diabetes Care, 2003, 26, 1589-1596, with the kind permission of the author).

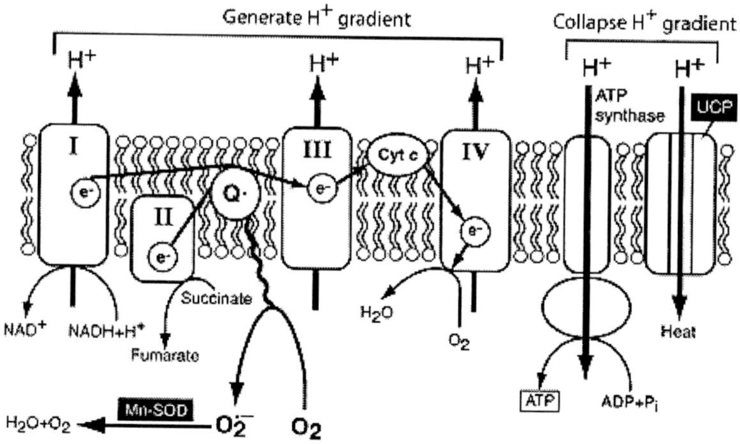

Figure 13. Hyperglycaemia-induced production of superoxide by the mitochondrial electron transport chain (published by Brownlee, M. The Pathobiology of Diabetic Complications. Diabetes, 2005, 54, 1615-1625, with the kind permission of the author).

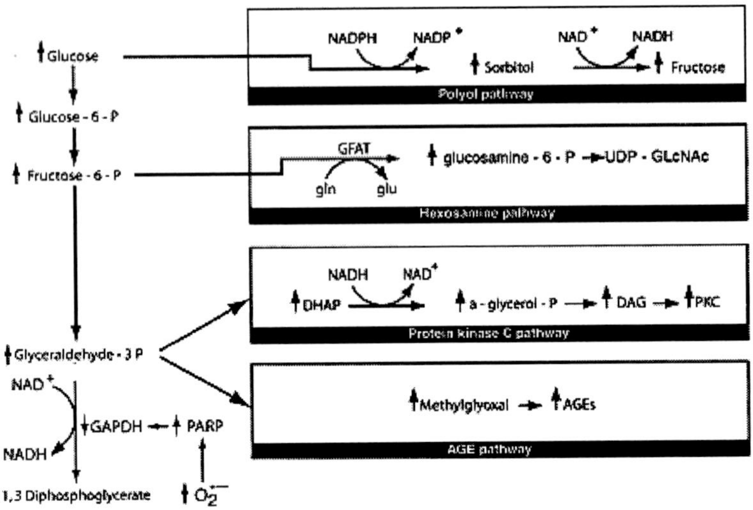

GAPDH - glyceraldehyde-3-phosphate dehydrogenase, GFAT – glutamine:fructose-6-phospate amidotransferase, GLN-glucosamine, glu-glutamine, UDP-uridine diphosphate, GlcNAc- N-acetylglucosamine, DHAP –dihydroxyacetone phosphate, DAG-diacylglycerol (activator of protein kinase C)

Figure 14. Mitochondrial overproduction of superoxide activates four major pathways of hyperglycaemic damage by inhibiting GAPDH (published by Brownlee, M. The Pathobiology of Diabetic Complications. Diabetes, 2005, 54, 1615-1625, with the kind permission of the author).

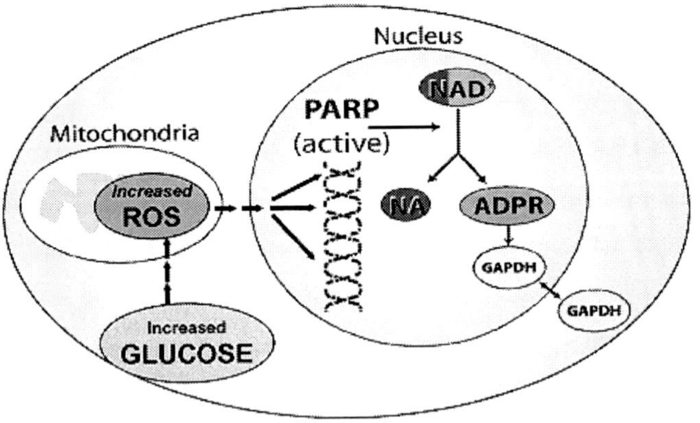

ADPR – ADP-ribose, NA- nicotinic acid, GAPDH- glyceraldehyde-3-phosphate dehydrogenase, PARP- poly(ADP-ribose)polymerase

Figure 15. ROS-induced DNA damage activates PARP and modifies GAPDH (published by Brownlee, M. The Pathobiology of Diabetic Complications. Diabetes, 2005, 54, 1615-1625, with the kind permission of the author).

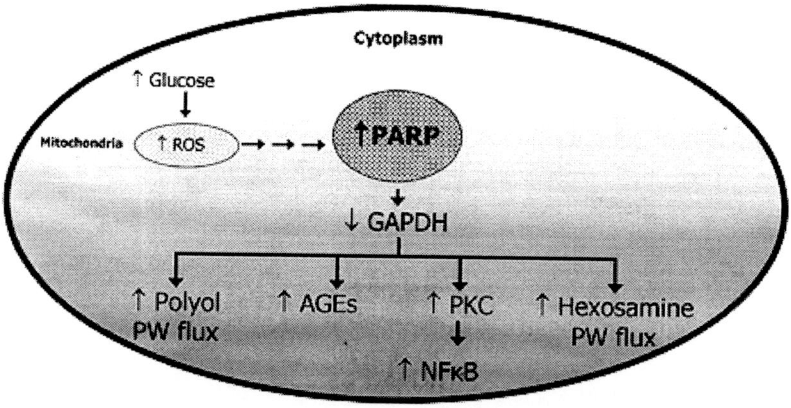

AGEs- advanced glycation endproducts, PKC – protein kinase C, NFκB- nuclear factor κB proinflammatory transcription factor, PW - pathway

Figure 16. The unifying mechanism of hyperglycaemia-induced cellular damage (published by Brownlee, M. The Pathobiology of Diabetic Complications. Diabetes, 2005, 54, 1615-1625, with the kind permission of the author).

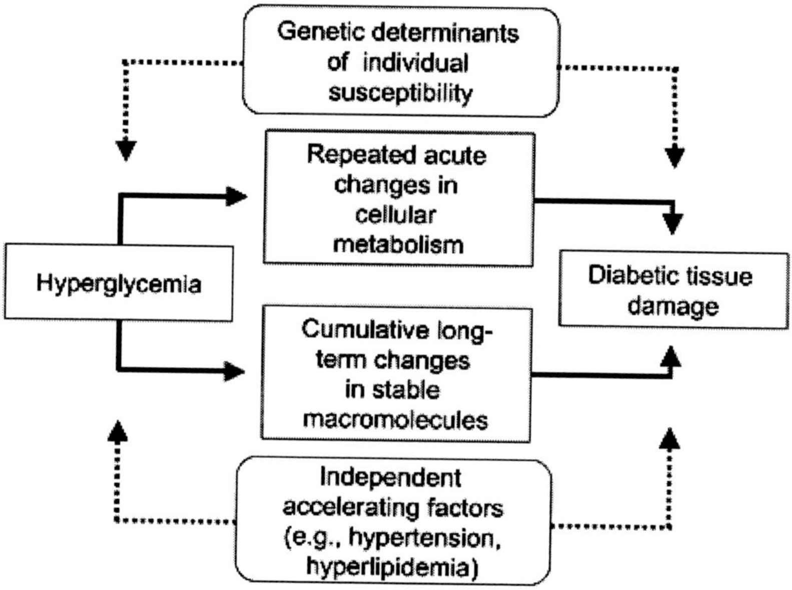

Figure 17. General features of hyperglycaemia-induced tissue damage. (published by Brownlee, M. The Pathobiology of Diabetic Complications. Diabetes, 2005, 54, 1615-1625, with the kind permission of the author).

Many previous studies analysed the influence of glucose on protein glycation and formation of advanced glycation end-products [108, 109, 110, 111, 112, 113, 114]. As shown for example by Kalousova and Skrha [115], accumulated glycation products (AGEs, but also advanced oxidation protein products – AOPP and advanced lipoperoxidation end products ALEs) exert not only direct effect on the extracellular matrix, but they are also bound to RAGE specific receptors. Beside other AGE-binding receptors, RAGE receptors are the best known and characterized as a 35kDa protein classified as a member of the immunoglobulin superfamily. RAGE can be expressed on the surface of various cell (e.g. monocytes, macrophages, mesangial cells, neurons, endothelial cells, smooth muscle cells and fibroblasts). It is a multiligand receptor and recognizes also β-amyloid, amphoterins and calgranulins involved in inflammatory processes. RAGE has also physiological function for the development of the central nervous system but during maturation, its presence decreases and is minimally expressed in adults. Under pathological conditions, such as diabetes melllitus, immune-inflammatory processes, Alzheimer´s disease, and cancer, RAGE expression is increased and binds AGE. AGE-RAGE interaction results in the activation of intracellular signal transduction pathways, such as MAPK and JNK kinases, activator protein 1 (AP-1) and nuclear factor NFκB. Subsequently, this activation is followed by overexpression of genes for cytokines (TNF-α, IL-1, IL-2, interferon γ, IL-8), growth factors (platelet-derived growth factor PDGF, insulin-like growth factor IGF-1), adhesion molecules (ICAM-1, VCAM-1). Other processes are also stimulated, among them stimulation of cell proliferation, increased vascular, migration of macrophages, formation of endothelin-1, increased synthesis of collagen IV, fibronectin and proteoglycans, increased procoagulant state [116, 117]. Little is known about the formation and accumulation of AGEs in young patients with T1DM. Tsukuhara [118] described increased urinary concentrations of pentosidine as AGE, and 8-hydroxy-2´-deoxyguanosine and acrolein-lysine as markers of OS in children with type 1 diabetes when compared with age-matched healthy control subjects. The mean age of 38 diabetic children was 12.8±4.5 years and their disease duration 5.7±4.3 years and HbA_{1c} 8.0±1.6%, 11 subjects had microalbuminuria >15 mg/g creatinine. The study indicated that the accumulation of AGEs is closely linked to oxidative stress and the subsequent endothelial dysfunction may start early in the course of type 1 diabetes. This means that the risk of vascular complications may be present at an early age and that the best possible glycemic control should be emphasized from the diagnosis of diabetes. Regarding these results, it is important to take into consideration also high and low haemoglobin glycation phenotypes in type 1 diabetes as presented by Hempe [119]. Different glycation phenotype are also associated with susceptibility to the glycation of other proteins, increased AGE synthesis, and greater risk of diabetes complications. Though there are analogies in hyperglycemia, oxidative stress and stimulation of consequent metabolic and inflammatory processes in different types of diabetes, T1DM and T2DM as the most frequent types of diabetes differ in their character. T1DM as autoimmune disease develops towards other autoimmunities and microvascular complications. T2DM (besides microvascular complications) inclines to atherosclerotic (macrovascular) complications, and to carcinogenesis [120, 121]. With regard to a variety of AGEs, there is also a large scale of AGE-inhibitors and AGE-breakers. These substances, either natural or synthetic, could display different reactions. They react with free amino groups of protein to modify them in order to prevent sugar attachment to protein, or they inactivate aldose and ketose sugars before reacting with amino groups of proteins. They could also be metal chelators, and antioxidants themselves like vitamin C or E. Another group are inhibitors trapping reacative dicarbonyl intermediates to form substituted triazines. There are Amadori adducts inhibitors, compounds inducing for enzymatic deglycation at the Amadori level. Finally, inhibitors called cross-link breakers are able to selectively cleave and break the established AGE-protein cross-links *in vitro* and *in vivo.* AGE-inhibitors and breakers for clinical use should be non-toxic, well absorbed across the gastrointestinal tract and achieve a broad distribution throughout various tissues. Their therapeutic impact can include not only diabetes, but also other diseases with AGE-formation and accumulation [114, 122].

Also the antioxidant defence is the focus of an intensive research in diabetes. The polymorphisms in the *Mn-SOD* and *EC-SOD* (CuZn-SOD) genes play certainly the role in the intensity of detoxifying superoxide. Chistyakov [123] found the relationship between Ala(-9)Val substitution in the *Mn-SOD* gene and diabetic neuropathy in a Russian population. The antioxidant gene enzyme expression and protein activity was studied on skin fibroblast cultures in type 1 diabetic patients with and without nephropathy, and nondiabetic control subjects. Skin fibroblasts of T1DM patients with diabetic nephropathy exposed to long-term hyperglycemia displayed unchanged expression and activity of Mn-SOD, catalase and GPx, whereas T1DM patients without complications, non-diabetic nephropathic patients, and control subjects had an intact stimulated antioxidant response to glucose-induced oxidative stress [124]. The present oxidative stress and low antioxidant defence was found not only in diabetic children, but similar tendency was observed also in their non-diabetic siblings, but not in their non-diabetic parents [125].

3.1.2. ROS, DNA Damage and DM

As mentioned at the beginning of this chapter, oxidative stress injures also DNA. It is impossible to enumerate all articles studying the intensity of oxidative stress and DNA damage in diabetic patients. Krapfenbauer [126] examined in 1998 type 1 and type 2 adult patients and compared them with age-matched non-diabetic controls. He confirmed the increased pentosidine and 8-hydroxy-2´-deoxyguanosine in urine in both diabetic groups and made conclusion that increased glycoxidation was approved but 8-OHdG could be taken either as consequence of increased DNA oxidation or reflection of DNA repair. Collins [127] used comet assay with endonuclease III to detect oxidized pyrimidines, and formamidopyrimidine glycosylase to detect damaged purines including 8-oxo-guanine. The increased formamidopyrimidine glycosylase-sensitive (FPG) sites were found in diabetic patients and the strong correlation was displayed between those sites and serum glucose concentrations. Another studies [128] summarized the importance of detection of 8-hydroxy-2´-deoxyguanosine (8-OHdG) in urine in different diseases. The methods for 8-OHdG detection and normal reference range for women and men were also discussed. Special attention was paid to 8-OHdG levels in patients with cancer , chronic inflammation, atherogenesis, and diabetes. An outline was made with regard to the levels of urinary 8-OHdG as oxidative stress marker and the presence of accelerated vascular dysfunction, and development of microvascular complications. The level of 8-OHdG could be also used as a marker of the effectiveness of dietary supplements with regard to whether they will reduce the oxidative damage. The review published by Samcova [129] described free radical different diseases and levels of 8OhdG. DNA damage and antioxidant defence were examined in peripheral leukocytes of T1DM patients by Dincer [130]. Comet assay was used in this study that confirmed increased DNA strand breakage and FPG-sensitive sites in diabetic patients. SOD and GPx activities in leukocytes were decreased. The presented findings showed increased oxidative DNA damage and impaired antioxidant defence in leukocytes of adult patients with T1DM. Similar findings of increased DNA damage in diabetic patients were confirmed in another study using comet assay [131].

3.1.3. ROS, DNA Repair and DM

The information about DNA repair process in relation to diabetes and oxidative DNA damage is scarce in humans. Attention is mostly paid to PARP and its inhibitors than to DNA repair process itself. PARP activates three major pathways of hyperglycemic damage in endothelial cells via inhibition of GAPDH activity (figure 16)[132].

Activity of PARP was found to be increased in whole retina and in endothelial cell and pericytes of diabetic rats. Long-term administration of PARP inhibitor (PJ-34) led to significantly inhibited death of retinal microvascular cell and prohibited from the development of early lesions of diabetic retinopathy in those animal models. In another experiment, PARP interaction directly with both subunits of NFκB (p50 and p65), mainly under hyperglycemic condition. The usage of PJ-34 blocked the PARP mediated activation of NFκB [133].

Japanese study published in 2004 [134] showed in a large group of T1DM patients that human *MTH1* gene Val83Met was associated with type 1 diabetes, particularly in females. The product of *MTH1* gene human mutT homologue 1 hydrolyzes 8-hydroxy-deoxyguanosine triphosphate (8-OHdGTP) to 8-hydroxy-deoxyguanosine monophosphate in order to prevent transversion mutation. MTH1 is localized not only in nucleus but also in the mitochondrial matrix assuring here the integrity of mitochondrial DNA. Val83Met polymorphism of *MTH1* gene leads to labile type of MTH1 and its decreased activity, both in nuclear and also in mitochondrial DNA. 8-OHdGTP can pair in DNA with cytosine or adenine and induce transversion mutations in mitochondrial DNA (mtDNA) as well as in nuclear DNA (nDNA). Such accumulated mutations would cause mitochondrial dysfunction, consequently more ROS production, leading to DNA strand breaks and the resultant recruitment of poly(ADP-ribose) polymerase for the repair.

Free-radical mediated ß-cell damage has been intensively studied in type 1 diabetes, but not in human type 2 diabetes. Oxidative DNA damage and intensity of DNA repair in pancreatic islets was assessed on pancreatic specimens (post mortem or post surgery) from patients with diabetes for 2-23 years and healthy controls. 8-OH-hydroxyguanine glycosylase (OGG1) DNA repair enzyme expression was examined using immunofluorescent staining of OGG1. The intensity and islet area stained for OGG1 displayed a significant increase in islets from T2DM subjects compared to the healthy controls. A correlation between increased OGG1 fluorescent staining intensity and duration of diabetes was also found. Most of the observed staining was cytoplasmic, suggesting that mitochondrial OGG1 accounts primarily for the increased OGG1 expression. The authors [135] concluded that oxidative stress related DNA damage may be a novel important factor in the pathogenesis of human type 2 diabetes and stimulates OGG1 islet expression. Blasiak [99] published in 2004 the results of study aimed at the intensity of DNA repair in patients with T2DM. Comet assay was used for the assessment of the level of endogenous basal DNA damage (DNA strand breaks and alkali labile sites), as well as endogenous oxidative and alkylative DNA damage (using DNA repair enzymes endonuclease III, formamidopyrimidine-DNA glycosylase and 3-methyladenine-DNA glycosylase II) in peripheral lymphocytes. Finally, the measurement of the sensitivity to DNA-damaging agents (hydrogen peroxide and doxorubicin) and the efficacy of removing this induced DNA damage in peripheral blood lymphocytes was performed. The levels of basal endogenous and oxidative DNA damage in diabetic patients were higher than in control subjects. There was no difference between the level of alkylative DNA in the patients and the controls. Diabetic patients displayed higher susceptibility to hydrogen peroxide and doxorubicin and decreased efficacy of repairing DNA damage induced by these agents than healthy control. The conclusion was made that patients with T2DM had the elevated level of oxidative DNA damage and the decreased efficacy of DNA repair and this may be the link between diabetes and cancer. At the same time, Varvarovska [136] published the study comparing the degree of DNA damage and the intensity of DNA repair in T1DM adults and children. Also in this case, comet assay was used for the evaluation of DNA damage and repair. Diabetic adults displayed higher DNA strand breaks and lower DNA repair capacity (measured in peripheral blood lymphocytes) than diabetic children. When adult patients were divided into 2 groups according the presence or absence of diabetic microvascular complications (DMC), the differences were more expressed, i.e. adults with DMC had higher DNA strand breaks and lower DNA repair capacity. The ratio between those two parameters was expressed as DNRI index that displayed large variations in adults without DMC and mainly in children.

3.1.4. Some Contributions to Antioxidant Therapy in Diabetes Mellitus

Advance in research of oxidative stress and developing knowledge of its impact on human metabolism has brought new therapeutic possibilities into experiments, clinical trials or clinical practice. In a review published by Schultz Johansen [137] there is a instructive summary of current antioxidant interventions. Several experiments on animal models were mentioned, among them usage of α-lipoic as antioxidant or SOD mimetic experiments in diabetic animals. Recently performed clinical trials with different antioxidants were enumerated. The HOPE trial was the largest trial conducted thus far for the use of antioxidants in diabetes, unfortunately no beneficial effect of vitamin E was approved on cardiovascular outcome or diabetic nephropathy. The Steno-2 trial compared the effect of a multifactorial intensive therapy with that of conventional treatment on modifiable risk factors for cardiovascular disease in T2DM patients. In the intensive treatment group, patients received combined pharmacotherapy (vitamin C, E, folic acid, chromium piconlinate and drugs lowering hyperglycemia, dyslipidemia and hypertension) plus behavior modification including low-fat diet, exercise and smoking cessation. Patients in intensive groups had a 50% decrease in the risk of cardiovascular events than a control group. Evidence is that a multifactorial approach is superior to conventional therapy for the prevention of oxidative stress-induced vascular complications in diabetes. Application of α-lipoic acid in the treatment of diabetic neuropathy had promising results and significantly improved patient symptoms were reached. A number of commonly used drugs have shown promising anitoxidant activity in addition to their primary pharmacological activity, such as thiazolidinediones, 3-hydroxy-3-methyl-glutaryl-Coenzyme A reductase (HMG-CoA) inhibitors (statins), and inhibitors of the renin-angiotensin system (inhibitors of angiotensin II activity, such as angiotensin converting enzyme inhibitors ACEIs and angiotensin II receptor blockers ARBs). In vitro and animal studies confirmed antioxidant effect of those drugs. These antioxidant properties may be a contributing factor to the therapeutic efficacy of these agents. Based on the new developments in the understanding of the pathophysiology of oxidative stress, new agents emerge and seems to be beneficial for the treatment of OS associated clinical problems (SOD and catalase mimetics, L-propionyl carnitine, PKC-β inhibitor). Another review published by Ceriello [106] summarized the choice of different antioxidant drugs and emphasized the importance of thiazolidinediones, statins, ACE inhibitors and AT1 inhibitors (figure 18). Da Ros summarized similarly causal antioxidant therapy with new promising tools influencing mitochondrial function and reducing DNA damage, blocking PKC, PARP and peroxynitrite [138, 139]. Reusch [140] reviewed the recent knowledge about diabetic complications and mechanism of glucotoxicity. It is emphasized that conventional antioxidants scavenge free radicals in an inefficient stoichiometric manner so that one molecule of the antioxidant would be neede to neutralize each free radical generated. Novel small mollecular weight compounds functioning as superoxide dismutase mimetics have more reliable benefits due to the catalytic properties that could permit enzymatic detoxification. Suggestion of PARP inhibitors is another therapeutic possibility for diabetic complication prevention. Li [141] pointed to the risk of long-term inhibition of PARP, e.g. premature aging, loss of genome stability. In animal experiment, low-dose PARP inhibitor-containing combination therapies (PARP inhibitor+vasodilators) showed beneficial effect on early peripheral diabetic neuropathy. Another approach how to face diabetic microvascular complications could be the application of aldose reductase inhibitors that block sorbitol formation [142].

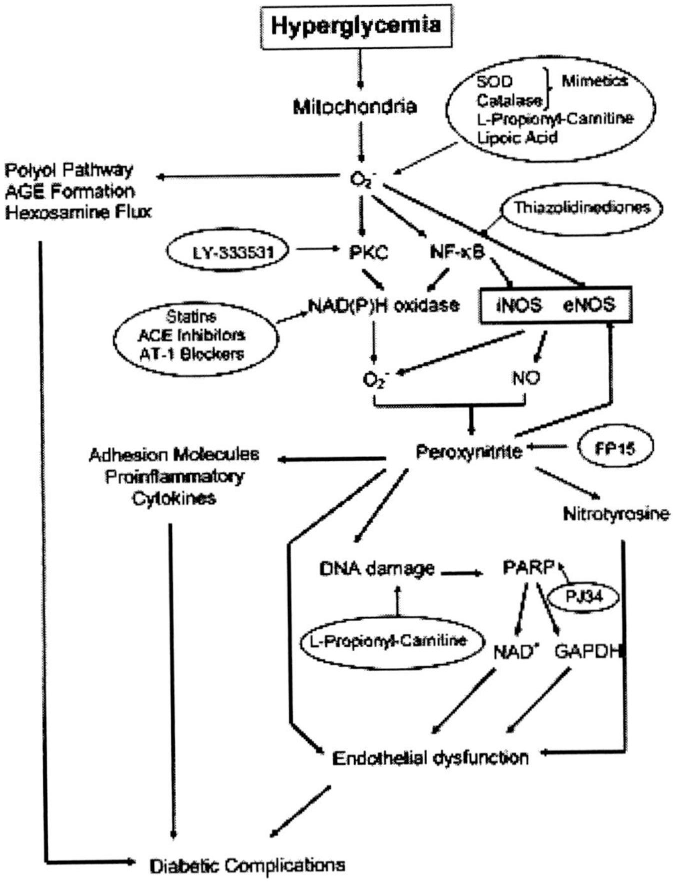

Superoxide overproduction reduces eNOS activity, but through NF-κB and protein kinase C (PKC) activates NAD(P)H and increases iNOS expression: the final effect is an increased NO generation. This condition favors the formation of the strong oxidant peroxynitrite, which in turn produces, in iNOS and eNOS, an uncoupled state, resulting in the production of superoxide rather than NO, and damages DNA. DNA damage is an obligatory stimulus for the activation of the nuclear enzyme poly(ADP-ribose) polymerase (PARP). Poly(ADP-ribose) polymerase activation in turn depletes the intracellular concentration of its substrate NAD^+, slowing the rate of glycolysis, electron transport, and ATP formation, and produces an ADP-ribosylation of the GAPDH. This process results in acute endothelial dysfunction in diabetic blood vessels, which contributes to the development of diabetic complications. NF-κB activation also induces a proinflammatory conditions and adhesion molecule overexpression. All these alterations produce the final picture of diabetic complications. SOD or catalase mimetics, L-propionyl-carnitine and lipoic acid, reducing mitochondrial overproduction of superoxide, and reducing DNA damage may be good candidates for causal intracellular antioxidants. This causal therapy would also be associated with LY 333531, PJ34, and FP15, which block protein kinase ß isoform, poly(ADP-ribose) polymerase, and peroxynitrite, respectively. Other current options may be thiazolinediones, statins, ACE inhibitors, and ATI inhibitors, which also have, at different levels, intracellular causal antioxidant activity. AGE, advanced glycosylation end product.

Figure 18. Location of action of different types of antioxidants in the treatment of diabetes mellitus (published by Ceriello, A. New insights on Oxidative Stress and Diabetic Complications May Lead to a "Causal" Antioxidant Therapy. Diabetes Care, 2003, 26, 1589-1596, with the kind permission of the author).

3.2. Crohn´s Disease

Crohn's disease (CD) belongs to idiopathic inflammatory bowel diseases (IBD) that also include idiopathic proctocolitis (IPC) and indeterminate colitis (IC), impossible to be ranked precisely into CD or IPC. These diseases cause the intestinal tissue to become inflamed, form sores and bleed easily. Symptoms include abdominal pain, cramping, fatigue, fever, diarrhoea sometimes bloody, malabsorption, loss of weight, failure to thrive in children. CD is characterized by an autoimmune granulomatous chronic process affecting any part of the gastrointestinal tract, from the mouth to the anus (figure 19). Patches of inflammation occur, with healthy tissue between the diseased areas [143](figure 20). The inflammation can extend through every layer of affected bowel tissue [143]. The generalized inflammation can also affect extragastrointestinal organs and their involvements are presented as skin affections (erythema nodosum, erythema multiforme), bone osteoporosis even osteonecrosis, joint inflammation under the picture of reactive arthritis or ankylosing spondylitis, liver affection as autoimmune hepatitis eventually primary sclerosing cholangitis, eye inflammation (episcleritis or anterior uveitis). Crohn's disease can not be cured by drugs or surgery, although either or both can help relieve symptoms [144].

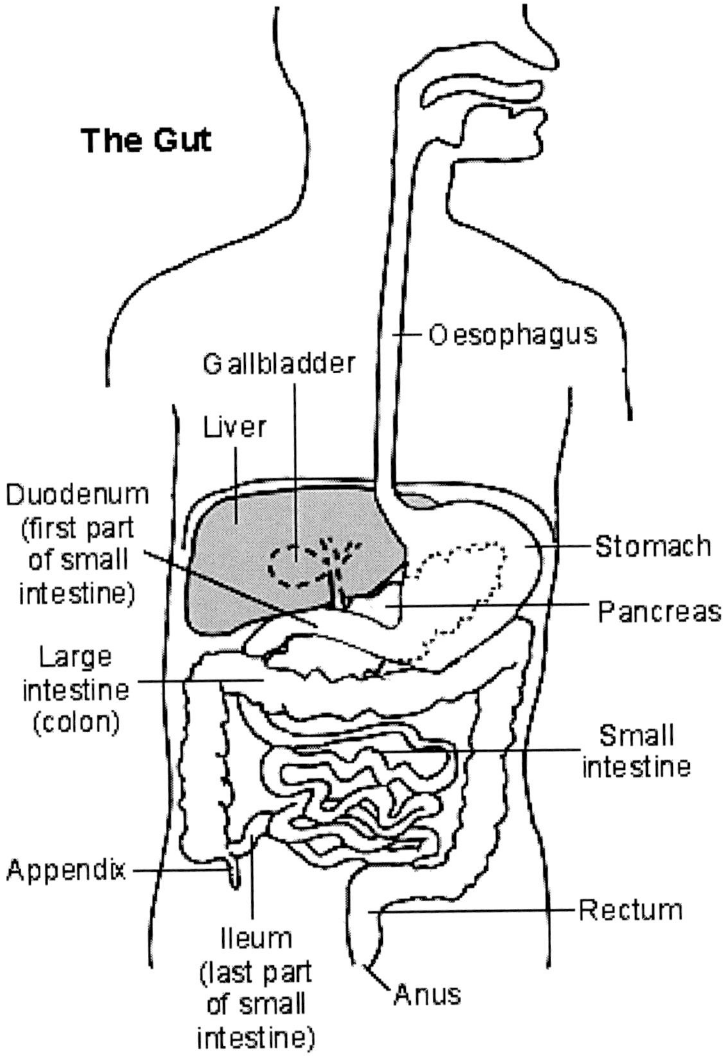

Figure 19. Distribution of inflammatory gastrointestinal affection in Crohn's disease.

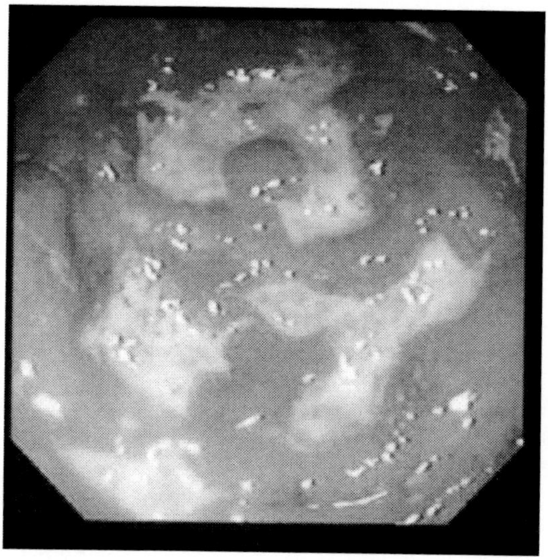

Figure 20. Inflammatory changes of gastrointestinal mucosa in Crohn's disease

Ulcerative colitis affects only the inner layer of the colon, or large bowel. It always starts in the rectum and may extend as a continuous inflammation from there into the rest of the colon. Usually ulcerative colitis can be controlled with medication. The disease can be completely eliminated by surgically removing the colon, but afterward, waste material may have to be stored and expelled through an external appliance [3, 144].

There is no known cause or cure for IBD. People are most frequently diagnosed between the ages of 15-25, or 45-55. IBD is particularly difficult for children and young adults since it often affects a person's self-concept, body image and lifestyle at a time when "being like everyone else" is so important. IBD is unpredictable. Most people experience periods of remission and flare-ups of the disease, often requiring long-term medication, hospitalisation or surgery [145]. Although IBD is found throughout the world, it seems to be more common in North America and northern Europe.

CD incidence keeps growing, and is high particularly in developed countries. In Europe, there is an incidence 9-10/100000 people per year, CD prevalence is about 20-80 persons per 100000 people. Both genders are affected equally.

Etiopathogenesis of CD is complex and includes genetic, immunological, infectious and other environmental factors. The formation of generalized granulomatous inflammation is accompanied by ROS formation [3, 146, 147, 148].

Initial inflammatory changes disrupt the protective intestinal barrier, then inflammatory cytokines are overproduced and inflammation progresses (cytokine phase). The next step includes tissue damage with increased inflammatory infiltration and simultaneous overproduction of ROS (figure 21). The further phase of healing is characterized by connective tissue formation in the intestinal wall and results in consequent stenoses and/or fistulae creation.

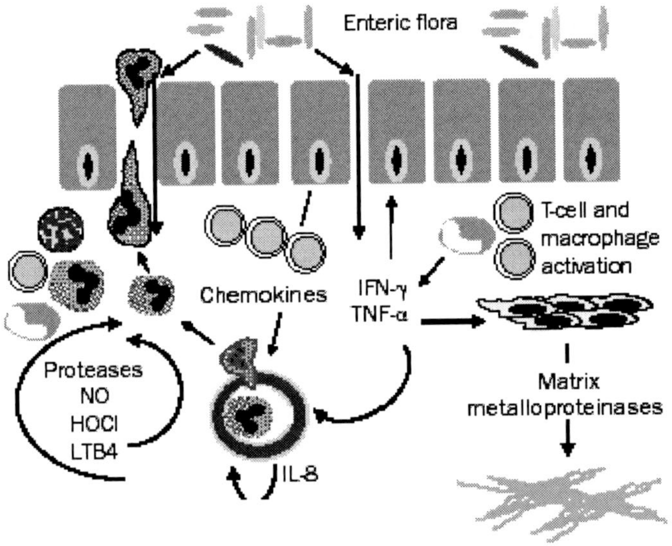

Figure 21. Immunological changes in the intestinal mucosa in CD (published by Shanahan, F. Crohn´s disease. The Lancet, 2002, 359, 62-69.

CD treatment is complex and encompasses anti-inflammatory (mesalazine), immunosuppressive (corticosteroids and azathioprine), and biological therapy with the aim to achieve the remission and to prevent disease complications. An appropriate nutritional support is indispensable, first total parenteral nutrition until the moment when the stabilisation of intestinal inflammation is achieved and the patient is able to stand the enteral nutrition without diarrhoea recurrence. The term of biological therapy is in CD reserved for the monoclonal anti-TNFα antibody that has success in refractory CD treatment. If conservative treatment in CD fails, the patient is indicated for surgical therapy with the management of Crohn's specific complications like stenosis, fistula, and abscess by bowel preservation. Limited resection and/or strictureplasty do not influence morbidity and rate and also time of recurrence [145].

3.2.1. ROS and IBD

As published long ago in 1991 by Yamada [149], the chronically inflamed intestine and/or colon is subjected to considerable oxidative stress whose source are the oxidants produced in phagocytic leukocytes (activated monocytes and polymorphonuclear leukocytes). These cells infiltrate in large numbers the intestinal mucosa and produce significant amounts of ROS in response to different inflammatory stimuli. Because the colonic mucosa contains relatively small amounts of antioxidant enzymes (SOD, catalase, GPx), the gut mucosa is overwhelmed during the time of active inflammation that results in intestinal injury [3, 149]. The complex review of oxidative stress as a pathogenic factor in inflammatory bowel disease brought Kruidenier in 2002 [146]. Here, the attention is paid to the origin of ROS and its impact on all cellular and extracellular components (figure 22). Evidence of increased reactive oxygen metabolite levels is mentioned as it was found in biopsy specimens of IPC or CD correlated with clinical and endoscopic indices of disease activity. The consequences of oxidative stress on the gut are analysed, such as increased lipid peroxidation, presence of the protein carbonyl content as well as peroxynitrite-modified proteins in colonic biopsies, also the presence of DNA oxidation expressed as increased presence of 8hydroxy-deoxyguanosine in the inflammatory changed mucosa, increased apoptosis in the epithelium. In addition, the importance of different antioxidant defences is emphasized. Non-enzymatic antioxidants represented by glutathione and metallothionein are discussed in detail, similarly much attention is devoted to antioxidant enzymes (SOD, GPx, catalase), their activities in IBD mucosa and their expression regulation. It has been confirmed that mainly Mn-SOD is highly inducible under inflammatory conditions as IBD are. Much less information is available concerning the expression regulation of catalase and GPx.. The last part of this review describes some possibilities of antioxidant therapy in IBD and its limitations [150). The other works by Kruidenier et al. [151] contributed to the better knowledge of an imbalance between oxidant and antioxidant mechanisms at the tissue level in patients with IBD. These studies focused on thorough systemic examination of intestinal resection specimens of patients with CD or IPC and compared with normal control mucosa. They confirmed different gut mucosal SOD isoform expression, without major differences between CD and IPC. A marked stepwise increase in Mn-SOD (mitochondrial) levels was observed in non-inflamed and inflamed IBD mucosae, but a proportion of this enzyme was found to be enzymatically inactive. The Cu/Zn-SOD (cytosolic) content decreased in epithelium with inflammation. Extracellular EC-SOD was found in low amounts in normal control mucosa, moreover with tendency in IBD patients to be decreased with inflammation. In lamina propria, SOD expression was much lower and not effected by inflammation. Mn-SOD and Cu/Zn-SOD were found mainly in neutrophils and macrophages, and EC-SOD was localized in small vessels, stromal cells, and neutrophils. These findings were extended with further study for the assessment of catalase, GPx, GSH, myeloperoxidase, and metallothionein (OH· hydroxyl radical scavenger) in the intestinal resection specimens of patients with CD or IPC versus normal mucosa. Both in CD and IPC patients, intestinal inflammation was associated with increased activities of catalase, GPx, and MPO, whereas the mucosal GSH content was unaffected. Despite this overall increase in mucosal H_2O_2 metabolising enzyme capacity, there were found the important differences between epithelium and lamina propria. The risk of direct hydrogen peroxide damage in the lamina propria seemed to be attenuated by the increasing numbers of catalase and MPO positive monocytes/macrophages and neutrophils that infiltrated the inflamed areas. But MPO overexpression might increase the lamina propria levels of hypochlorous acid, a stable ROS with multiple pro-inflammatory effects. Epithelial cells expressing catalase and GPx were found unchanged during inflammation, and moreover hydroxyl radical scavenger metallothionein displayed decreased expression, thus epithelial antioxidant defence is low. Another study performed by Kruidenier et al. confirmed increased levels of peroxynitrite in the inflamed mucosa of IBD patients, and also elevated malondialdehyde concentration as a marker of lipoperoxidation [147]. All these results confirm an imbalanced and inefficient endogenous antioxidant response to ROS produced in the inflamed intestinal mucosa of IBD patients and pointed out the importance of oxidative stress in the pathogenesis of IBD [146, 147, 150, 151]. In another study [152], patients with CD had increased markers of lipid peroxidation as measured by plasma F_2 isoprostane, breath pentane and ethane output and significantly lower plasma antioxidant vitamins (ascorbic acid, α- and β-carotene, lycopene, and β-cryptoxanthin) when compared with control subjects. High level of GPx were found in a Turkish group of IBD patients indicating in this way stimulated response against oxidative stress. MDA levels in this study were normal in IBD individuals and this finding suggests the clearance of free radicals without leading to important lipid peroxidation [148]. A large group of IBD patients (94 with IPC, 97 with CD) was compared with 72 healthy controls with regard to blood total antioxidant capacity TAC and corrected TAC cTAC (TAC after subtraction of the interactions due to endogenous uric acid, bilirubin, and albumin). In both groups of IBD, TAC and cTAC were significantly reduced when compared with controls irrespective of disease activity [153]. The study on animal mice model with dextran sulphate sodium (DSS)-induced colitis used the preventive (before colitis induction) or treatment (after colitis induction) with GSH or CoQ. Colitis symptoms were reduced in both groups of animals, hence GSH or CoQ had a beneficial effect on acute signs of IBD. Controversy brought into the problem the histological evaluation that revealed increases in colonic dysplasia and ulceration after GSH or CoQ supplementation [154]. The more favourable result had another animal study with DSS induced colitis. Mice treated with three different antioxidants (S-adenosylmethionine, a GSH precursor; green tea polyphenols; a cysteine prodrug - 2(R,S)-n-propylthiazolidine-4(R)-carboxylic acid – PTCA). All the antioxidants significantly improved diarrhoea and colon lesions, in blood reduced GSH levels returned to normal when treated with antioxidants. Serum amyloid as acute phase protein and tumour necrosis factor-alpha were significantly increased in animals with florid colitis, but they improved with antioxidant treatment. These results may support a beneficial role of antioxidant therapy in IBD individuals [155]. Another recent study used the treatment with N-acetylcysteine (NAC) against toxic damage in the rat colitis induced with acetic acid. NAC was applied locally (into colon) and systemically (intraperitoneally). Treatment with higher levels of NAC (100mg/kg for 7 days) significantly decreased the degree of colonic injury and NAC is judged as having a role in the treatment of IBD [156]. Italian authors published an article on IBD patients and their plasma vitamin A, E, and carotenoid concentrations. They found significantly reduced antioxidant levels in IBD patients (both IPC and CD), particularly in those with active disease, when compared with healthy controls. 8-hydroxy-deoxy-guanosine (8-OHdG) were also assessed in the patients´leucocytes. A significantly increased levels of 8-OHdG were found without regarding the disease activity and antioxidant levels [157].

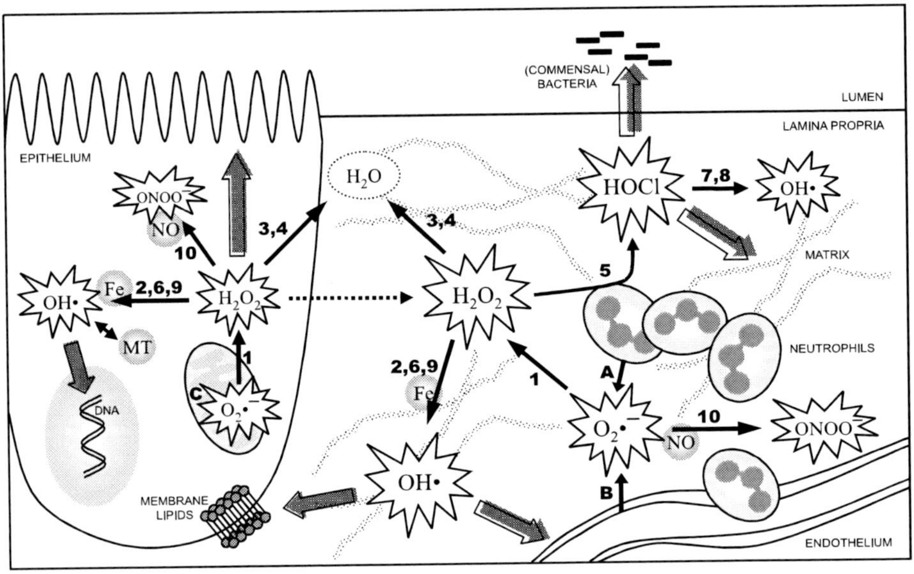

A. - neutrophil NADPHoxidase, B-xanthine oxidase, C-mitochondrial NADPH cytochrome p450 reductase

1-formation of hydrogen peroxide from superoxide, 2-formation of OH and OH- from hydrogen peroxide (Fenton reaction)

3-formation of oxygen and water from hydrogen peroxide catalysed by catalase or GPx

4-formation of water nad NADP+ from hydrogen peroxide, H+ and NADPH

5-formation of hypochlorous acid from hydrogen peroxide and chlorine (catalysing enzyme myeloperoxidase)

6-formation of O_2, OH·and OH- from hydrogen peroxide and superoxide (iron-catalysed Haber-Weiss reaction)

7-formation of oxygen, chlorine and hydroxyl radical from hypochlorous acid and superoxide

8-formation of OH·and OH- and oxidized iron from reduced iron ions and hypochlorous acid

9-formation of nitrogen dioxide, hydroxyl radical and H+

10-formation of peroxynitrite anion and H+ from nitric oxide and superoxide

No known enzyme exists to facilitate the detoxification of hydroxyl radical OH· , OH·-induced tissue damage may be prevented by the binding of transition metal ions by albumin,k ceruloplasmin, ferritin, transferrin and metallothionein.

Figure 22. Schematic representation of ROS in intestinal inflammation (published by Kruidenier, L; Verspaget, HW. Oxidative stress as a pathogenic factor in inflammatory bowel disease — radicals or ridiculous? Alimentary PharmacologyandTherapeutics, 2002, 16, 1997-2015, with the kind permission of the author)

3.2.2. ROS, DNA Damage, DNA Repair and IBD

An association between IPC and an elevated risk for colorectal cancer is well established and this carcinogenesis is probably driven by chronic inflammation. The available evidence suggests that DNA damage caused by oxidative stress in the characteristic damage-regeneration cycle is a major contributor to colorectal cancer development in IPC patients [158]. Oxidative DNA damage (8-OhdG) in mucosa of ulcerative colitis has been examined in the D´Inca study [159]. Patients with IPC and dysplasia had significantly higher mucosal 8-OhdG concentrations, in particular in older patients, in patients with long-standing disease, and active disease when compared with sporadic colon cancer and irritable bowel syndrome. Also the findings in patients with CD demonstrated increased levels of 8-OhdG, this time in peripheral leucocytes [D´Odorico]. An older study on a small number of patients with IPC [160] showed deficient repair of radiation-induced DNA damage compared with a group of healthy controls. The development of colorectal cancer is the most serious long-term complication faced by patients with longstanding extensive ulcerative colitis, with an incidence 20-fold higher than colorectal carcinoma (CRC) in the general population. Thus, more attention is paid to the intensity of DNA repair, expression of DNA repair genes (BER or MMR), and their association with IPC [161,162, 163]. The proper course of the cell cycle is controlled by protein p53 that is overexpressed in the G1 stage of the cell cycle when DNA damage is present. In order to allow DNA repair in stressed or injured cells, protein p53 blocks the continuation of the cell cycle and DNA damage can be repaired. When the expression of p53 was measured in inflamed and regenerated mucosa of patients with IBD, the increased number of cells expressing p53 was found in patients with IPC than with CD. This finding supports the idea of importantly changed DNA and the cell cycle arrest mediated through p53 [164]. Oxidative stress induces DNA strand breaks and their accumulation leads to the stimulation of poly(ADP-ribose)polymerase (PARP), that further initiates BER pathway, but also stimulates pro-inflammatory pathways as mentioned previously. The study on IL-10 deficient animal models showed, that the application of PARP inhibitor (3-aminobenzamide) led to the normalization of colonic permeability and downregulation of proinflammatory cytokines, such as tumour necrosis factor-α and interferon [73].

The information about the intensity of DNA repair in CD patients is lacking.

Oxidative stress undoubtedly contributes to the pathogenesis of both chronic lifelong diseases - diabetes mellitus and Crohn's disease though the origin of oxidative stress differs. In DM, oxidative stress is mostly of metabolic origin, in CD, oxidative stress is in particular of inflammatory origin. The extensive research of both diseases has brought recently an enormous knowledge of molecular genetics and participating genes, immunological processes and their consequences, metabolic changes with special attention to oxidative stress. What remains unpredictable is the future course of either diabetes or Crohn's disease in newly diagnosed patients.

Chapter IV

Our Study Examining Oxidative Stress, DNA Damage and DNA Repair in Patients with Diabetes Mellitus and Crohn's Disease

Many previously published reports established the fact of increased oxidative stress in patients with diabetes mellitus and inflammatory bowel diseases (IBD), in particular Crohn's disease (CD). Also our previously published article (JDC) reported increased parameters of oxidative stress in children with type 1 diabetes mellitus (T1DM), and similarly their non-diabetic siblings, when compared with age-matched healthy children. Though the finding of increased oxidative DNA damage in diabetic individuals as well as in subjects suffering from IBD has been well documented in several studies, the information about the intensity of DNA repair in these patients is completely lacking.

The acknowledged existence of interindividual differences in DNA repair [165] and our acquaintance with a well established method comet assay (single cell gel electrophoresis used for the assessment of DNA strand breaks and in a modified version for the measurement of DNA repair capacity) [65] initiated our hypothesis (figure 23).

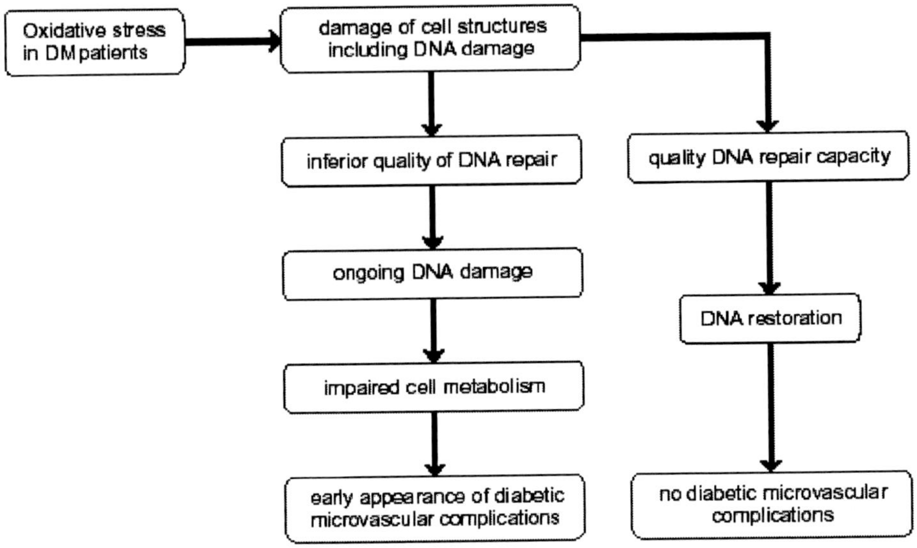

Figure 23. Hypothesis of influence of DNA repair on the course of DM or IBD; (in DM diabetic microvascular complications and macrovascular complications, in IBD patients the presence of fistulae, stenosis, abscess)

4.1. Hypothesis

Summarizing the hypothesis: these patients with free radical diseases, such as diabetes or IBD, that would have low DNA repair capacity could be expected as those having an early development of complicated course of disease and vice versa, those ones having good or excellent DNA repair capacity should not have an early development of disease complications.

In order to get the basic information about the intensity of DNA repair capacity in patients with diabetes mellitus and Crohn's disease, adult and paediatric patients with type 1 diabetes mellitus, adults with type 2 diabetes, adults and children with Crohn's disease were recruited and examined for the intensity of oxidative stress (parameters of OS) and for oxidative DNA damage (DNA strand breaks) and DNA repair capacity (DNArc).

4.2. Study Material

T1 diabetes mellitus patients: 53 adult patients, 28 women, 25 men, mean age 38.25±11.02 years, mean disease duration 16.49±3.22 years, mean HbA1c concentration 8.36±1.12 % (normal ranges 4.3-5.8% DCCT, performed 2002) were enrolled in the study. Among them 23 persons without any diabetic microvascular complication (DMC), 30 remaining persons having 1 or more diabetic complications - maximum 4 complications (15 subjects): diabetic retinopathy (18 subjects), diabetic nephropathy (22 subjects) diabetic neuropathy (11 subjects) and diabetic foot (3 patients). Diabetic retinopathy was diagnosed by a specialized ophthalmologist, diabetic nephropathy was defined as microalbuminuria higher than 20μg/min and/or proteinuria higher than 0.20g/24h, diabetic neuropathy was examined using a neurological disability score (NDS), electromyography (EMG), biothesiometer and monofilamenta and diabetic foot was evaluated according to International Consensus on the Diabetic Foot. 5 patients with DMC had simultaneous macrovascular complications (coronary heart disease, or peripheral vascular disease, or cerebral ischaemia).

These 2 groups of patients (without complications and with complications) did not substantially differ in age, sex, and diabetes control, but they differ in disease duration (14.70±3.11 years x 17.63±2.57 years, $p<0.01$). All patients were on diet, and treated with intensified insulin therapy or on insulin pump, other medicaments were applied according to the present diabetic complications.

Type 2 diabetes adults (T2DM): 14 patients with T2DM, 6 women, 8 men, mean age 58.0±5.6 years, mean disease duration 11±6.06 years, HbA1c 9.9±2.5% DCCT, treated with insulin, and/or PAD, + diet. 5 patients had macrovascular complications (cardiac artery disease or peripheral vascular disease), 5 patients had microvascular complications (retinopathy in 4 cases, once nephropathy, neuropathy in 3 patients), 3 patients had both types of complications.

Crohn's disease patients: 17 adult patients, 11 women, 6 men, mean age 37±11.36 years, mean disease duration 10.53±7.24 years, CDAI index 68.18±41.10 (normal ranges up to 150 as inactive disease, above 150 active disease) were examined. 11 patients underwent one or more operations for CD complications (stenoses, fistulae), all subjects were on intensive treatment including salicylates, corticosteroids and/or immunosuppressive drugs (Imuran). 1 patient was treated with biologicals (Remicade).

T1DM children: 20 children, all individuals were without diabetic microvascular complications, 8 girls and 12 boys, mean age 13.3±2.9 years, mean disease duration 2.94±2.72, mean HbA1c 9.50±2.52% (DCCT) (study performed in 2002-2003) were enrolled. All children were on diet and treated with intensified insulin therapy or on insulin pump.

17 children with Crohn's disease: 7 girls, 10 boys, mean age 14.29±SD 2.52 years, mean disease duration 3±1.43 years, PCDAI index (activity of disease) was 35.69±16.60 (normal up to 30), CDAI index (activity of disease compared with adults) was 112.94±82,7 (normal up to 150). All patients were treated with salicylates, and/or corticosteroids, at the time of blood collection, none, at the time of blood sample collection, was treated with biologicals. 3 patients underwent surgical treatment for stenosis, fistula, and abscess complications.

Control group of healthy adults: healthy control subjects were previously recruited by the corresponding laboratories and their results served as normals for appropriate methods (for markers of oxidative stress 46 women and 42 men, total mean age 34.80±10.32 years, for DNA strand breaks and DNA repair capacity 20 women and 22 men, mean age 45.20±8.31 years).

Control group consisted of 11 healthy children randomly chosen among children examined for accidental heart murmur (heart defect excluded using echocardiography) 5 girls, 6 boys, mean age 13.7±3.8 years.

T1DM and CD children were comparable with regard to their age and sex with the control group of children.

4.3. Methods

Methods used for the patients´ assessment: blood samples were examined for markers of oxidative stress, such as superoxide dismutase SOD in erythrocytes (U/g Hb), glutathione peroxidase GPx in blood (U/g Hb), antioxidant plasmatic capacity AOC (mmol/l), reduced glutathione GSH in erythrocytes (mmol/l ery), plasmatic malondialdehyde MDA (μmol/l).Kits from Randox (Crumlin, Great Britain), adapted to an automated Hitachi 717 analyser (Roche, Mannheim, Germany) were used for determination of superoxide dismutase in erythrocytes, glutathione peroxidase in whole blood, and finally for plasma antioxidant capacity. Reduced glutathione concentration in erythrocytes was estimated with kits from Oxis Int. (Bonnenie, Marue, France) while plasma malondialdehyde as a product of lipid peroxidation was determined photometrically as TBARS (thiobarbituric acid reactive substances) [125, 166].

DNA damage and DNA repair capacity in patient's peripheral lymphocytes were examined with comet assay (single cell gel electrophoresis) and its modified version (figures 24, 25, 26).

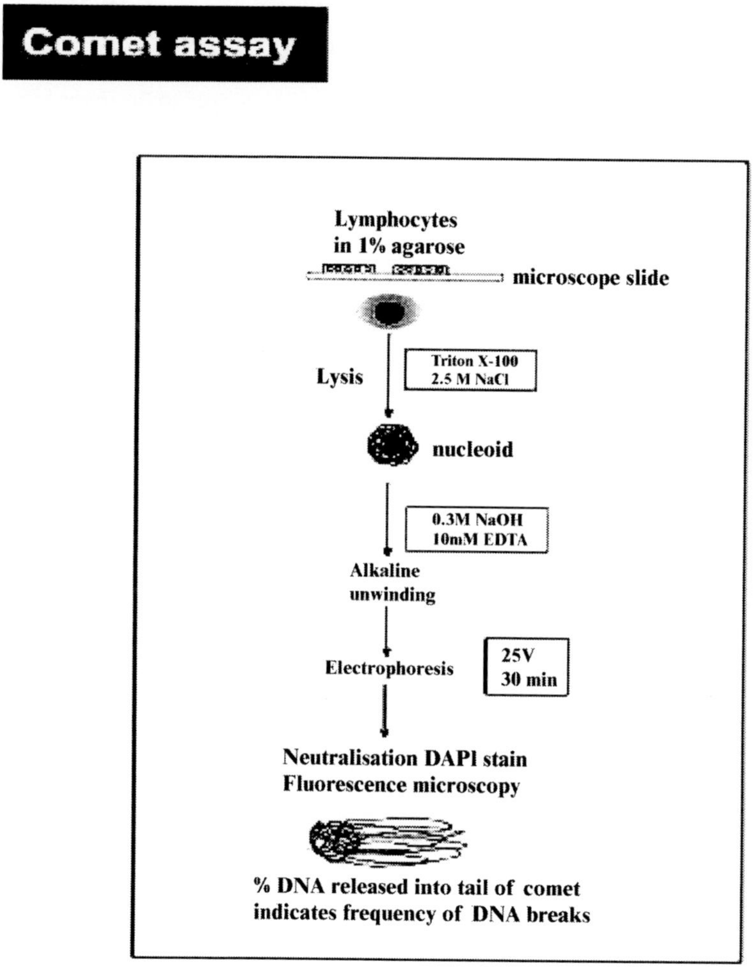

Figure 24. Comet assay for the evaluation of DNA strand breaks (by A.R. Collins).

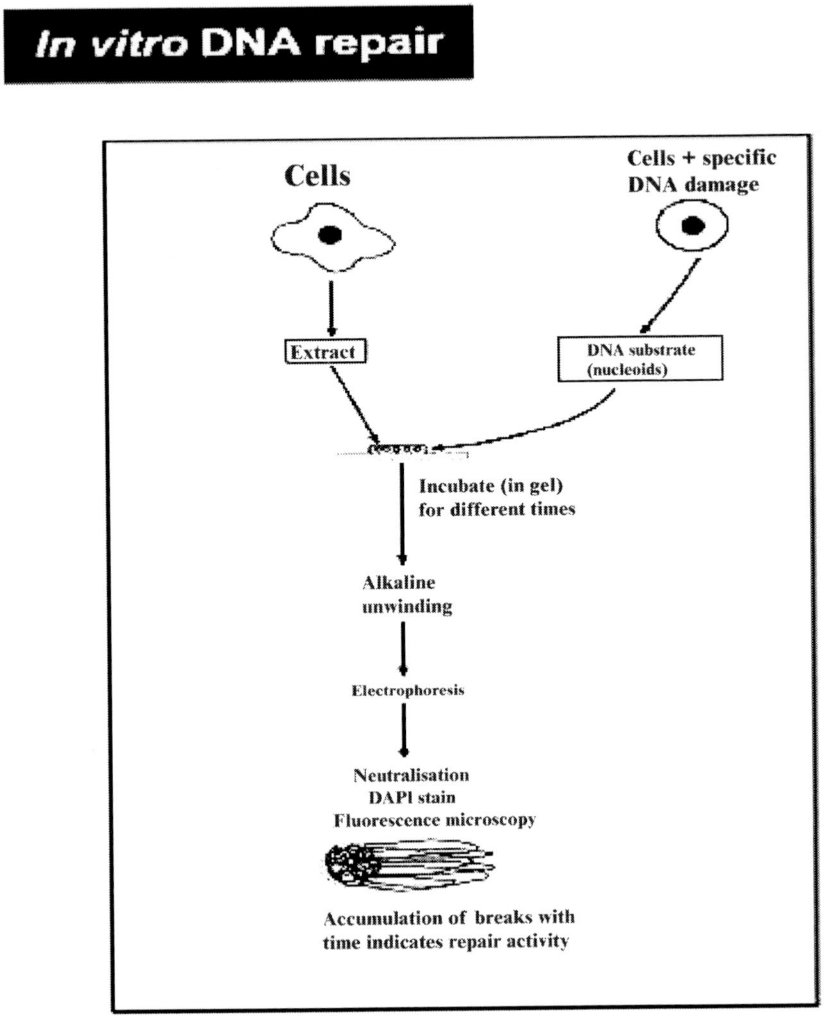

Figure 25. Modified comet assay for the evaluation of DNA repair (by A.R. Collins).

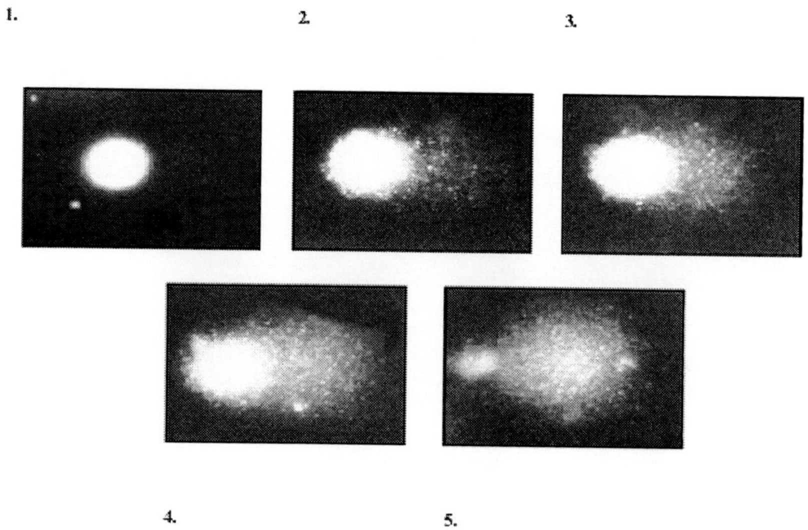

Figure 26. Comet assay - different degrees of DNA damage (1. nucleoid without damage, 3. nucleoid with comet and present DNA breaks 5. completely destroyed nucleoid (published by AR Collins, The comet assay for DNA damage and repair, Molecular Biotechnology, 2004, 26, 249-261, with kind permission of the author).

DNA strand breaks (DNAsb) measured by X-ray-calibrated comet assay (alkaline single-cell gel electrophoresis) [65], in peripheral blood lymphocytes isolated from heparinized blood samples were used as a parameter of oxidative DNA damage. Results are expressed as strand breaks per 10^9 Daltons of the DNA. Briefly, cells embedded in a thin layer of agarose on a microscope slide are lysed with detergent and high salt, and the resulting nucleoids electrophoresed at high pH. DNA is pulled out of the nucleoid to form a "comet tail". Comets are visualized by fluorescence microscopy, after staining with ethidiumbroide. The relative amount of DNA in the tail compared with the head reflects the number of DNA breaks present.

The individual DNA repair capacity (DNArc) restoring oxidative DNA damage was measured by modified comet assay using cell extracts from patients´ lymphocytes. In this case, cultured cells (HeLa) treated with a photosensitiser Ro 19-8022 Hoffman-La Roche and visible light are used as a target cells with a defined amount of oxidative damage. These cells embedded in agarose are treated for 45 min. with cell extracts prepared from lymphocytes isolated from tested patients. The number of DNA breaks (expressed as strand breaks per 10^9 Daltons) formed in the 8-oxoguanine-containing DNA of target cells by cell extracts expresses the capacity of the examined lymphocytes to initiate the repair of oxidised bases (8-oxoguanines) [65, 167].

To express the relationship between the intensity of individual DNA repair capacity and the existing level of oxidative DNA damage, a *DNA repair index* (DNRI) where DNArc represents the number of strand breaks/10^9 Daltons induced by cell extract in DNA of target (HeLa) cells containing 8-oxoguanine and DNAsb is the number of strand breaks/10^9 Daltons measured in DNA of tested lymphocytes [136].

Statistical evaluation: data were expressed as means±standard deviation. Wilcoxon unpaired test, Spearman and Kruskal-Wallis test and rank correlation were used in statistical evaluation of the results. Statistical significance was implied by p-value <0.05.

The Local Ethics Committee approved the study, and patients´ informed consent was obtained from all participants before entering the study.

4.4. Results

All the graphs are expressed in % of values. One group, usually healthy adults or healthy children, were indicated as a baseline 100% value and statistical average and standard deviation of all other groups were expressed as the percentage of the baseline value.

Markers of oxidative stress includes superoxide dismutase SOD, glutathione peroxidase GPx, antioxidant plasmatic capacity AOC, reduced glutathione GSH, malondialdehyde MDA.

Evaluation of DNA damage is indicated as DNAsb. DNA repair capacity as DNArc. Statistical significance $^{*}p<0.05$ $^{**}p<0.01$ $^{***}p<0.001$

If standard deviation surpasses the size of the graph, its value is mentioned below the graph.

Axis X - examined parameters axis Y - % of values

This pilot study brings some new information about the intensity of DNA damage and DNA repair capacity in seven different groups of individuals, among them five groups of patients with chronic lifelong diseases and still among them, two groups of pediatric children. As we would like to get some knowledge about the development of the assessed parameters not only with regard to the patients´ diagnosis, but also with regard to the patients´age (having in mind the theory of aging), we dared to compare the results of diseased children with diseased adults, and also to relate all the results to a baseline value of healthy adults. Concerning children, we accomplished the comparison between both groups of pediatric patients with healthy controls.

4.4.1. Study of Oxidative Stress, DNA Damage and DNA Repair Capacity of Lymphocytes in Healthy Adults and Healthy Children

The level of SOD was significantly higher in healthy children than in adults ($p<0.001$), on the other hand GPx was lower ($p<0.001$), the remaining parameters of oxidative stress (AOC, GSH, MDA) did not differ substantially. DNA strand breaks were comparable in both groups (a large variance of results in children influences the statistical significance), but DNA repair capacity was significantly higher in children ($p<0.001$) together with DNRI ($p<0.001$).

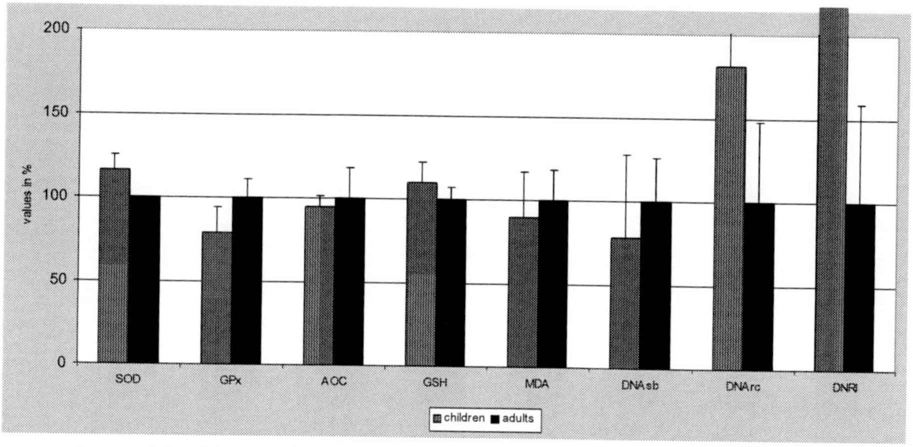

mean values of healthy adults=100%, DNArc SD in healthy children SD±36%, mean DNRI in healthy children 1044% of the value of healthy adults, DNRI SD in healthy children SD 3026% of the value of healthy adults

Figure 27. Markers of oxidative stress, DNA damage and repair.

The activities of antioxidant enzymes and levels of other antioxidant substances do not change substantially with age. Our group of healthy children had higher SOD and lower GPx when compared with healthy adults. This result corresponds with the finding of SOD and catalase activation in physically active children [168]. Similar activation of catalase and glutathione reductase and reversely decrease glutathione peroxidase in healthy men after an extreme cycling performance was found by Aguilo [169]. It could be expected that our healthy children had certainly more physical activity than healthy adults. Oxidative damage of healthy children was lower than in adults, but their DNA repair capacity and DNRI were significantly higher than in healthy adults. This finding certainly correlates with age difference between children and adults, and thus age-related decreased activity of DNA repair enzymes, as published by several authors and implied in the theory aging [83, 84, 170, 171, 172]. Raji [173] confirmed an increased DNA repair in children. Chen [174] found age-associated decrease of oxidative repair enzymes, namely human 8-oxoguanine DNA glycosylase.

4.4.2. Comparison of T1DM Patients Versus Healthy Population

T1DM patients showed lower GPx (both groups $p<0.01$) and lower antioxidant plasmatic capacity (both groups $p<0.01$). Their MDA was very high (both T1DM groups $p<0.001$). SOD was significantly high in T2DM adults ($p<0.001$), on the other hand GSH as the important intracellular antioxidant was significantly decreased ($p<0.01$) in the same group of patients. Their MDA was substantially low ($p<0.01$). Metabolic compensation (evaluated as the value of HbA1c) of T1DM adults was significantly better (HbA1c 8.36±1.12 % DCCT) than this one of T2DM adults (HbA1c 9.9±2.5% DCCT, $p<0.05$) and T1DM children (HbA1c 9.50±2.52% DCCT) ($p<0.05$), (it is not shown in the figure).

The findings of lower GPx and AOC and high MDA both in adults and children with T1DM correspond with the results of other authors [130, 175, 176]. Results of patients with T2DM showed stimulation of antioxidant enzymes and low GSH. Low MDA could be explained by MDA incorporation into LDL that are usually elevated in T2DM [177].

The results of T1DM patients behaved differently. Significantly increased DNA repair capacity was found in both adult and pediatric groups of T1DM individuals when compared with adult healthy controls ($p<0.001$ both), and along with it, DNRI was also extremely elevated (both groups $p<0.001$).

There is an apparent finding of significantly elevated DNA strand breaks in patients with T2 diabetes ($p<0.05$). On the other hand, their DNA repair capacity showed an extremely low level of repairing strand breaks ($p<0.001$) and similarly DNRI is extremely low ($p<0.001$).

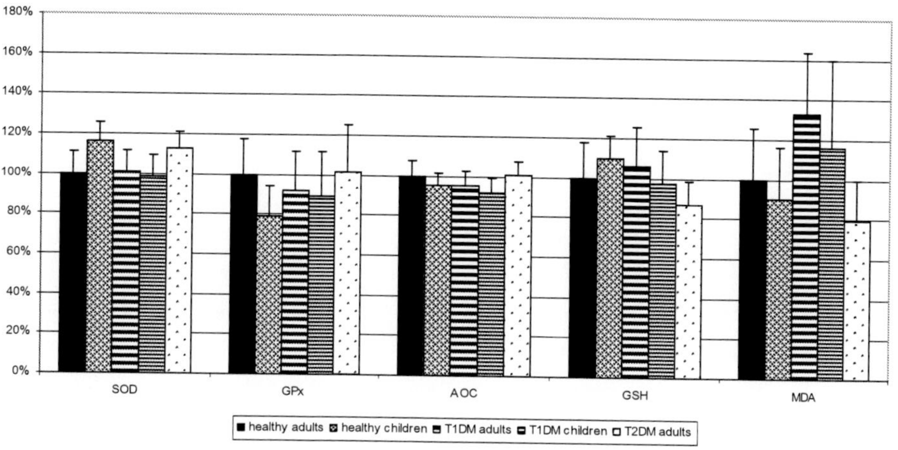

mean values of healthy adults=100%

Figure 28. Markers of oxidative stress.

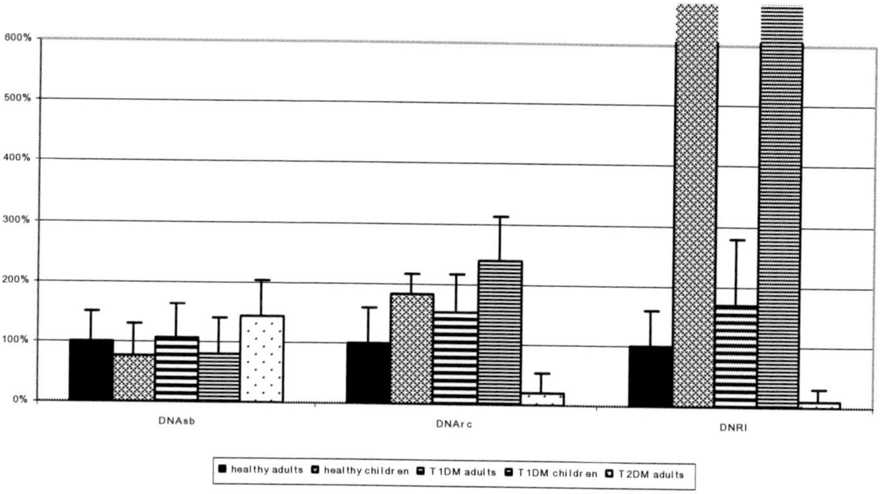

mean values of healthy adults=100%, DNRI in healthy children mean 1144% SD 3026%, T1DM children DNRI mean 919% SD 1307%

Figure 29. DNA damage and repair.

Several studies confirmed oxidative DNA damage in diabetic patients using different methods, mainly with increased loss of oxidized bases into urine or an increased number of DNA strand breaks [3, 126, 127, 129]. In our study, the differences in DNA damage between healthy controls and T1DM (adults, children) were not statistically significant. T2DM patients had significantly elevated DNA strand breaks. Similar results were described by Sardas [131] who found worse DNA damage in T2DM than in T1DM. Krapfenbauer [126] also confirmed increased DNA oxidation in both types of diabetes. On the other hand, our T2DM patients had a very low DNA repair. It could be explained either by the older age of this group, or the problem could be hypothetically also in the genetically or environmentally influenced bad DNA repair. Similar results were presented by Blasiak et al.[99] in a case-control study and he found also low DNA repair in T2DM subjects. This fact could represent a link to susceptibility of T2DM patients to carcinogenesis. Generally, low DNA repair can lead to extreme DNA damage, and thus initiate apoptosis, or inaccurate repair of damaged DNA leads to mutations and genomic instability and finally to carcinogenesis [2, 3, 53, 55, 56, 57].

4.4.3. Comparison of T1DM Adult Group Versus T1DM Children

This graph matches results of diabetic children and adults with T1DM. Though a worse diabetes compensation in children measured as glycated hemoglobin HbA_{1c} (significantly higher, $p<0.05$) and lower levels of AOC ($p<0.05$), diabetic children had favourable low levels of MDA ($p<0.05$), i.e. low lipid peroxidation. They showed also significantly lower DNA damage ($p<0.05$) and stimulated DNA repair ($p<0.001$), as well as DNRI (also $p<0.001$).

The presented results cannot be compared with a similar study in T1DM children and adults. Nevertheless, it is obvious that young organisms coped better with the requirements of hyperglycemic state and oxidative load. The influence of a shorter disease duration (2.94±2.72 years in children versus 16.49±3.22 years in adults, $p<0.001$) could also influence the presented results. Genetically conditioned individual DNA repair of oxidative damage (via BER) might also affect the results. Unfortunately, the matched results of DNAsb, DNArc and DNRI measured in each patient after a 2-year interval are not yet available, hence it is impossible to declare an attitude towards this question.

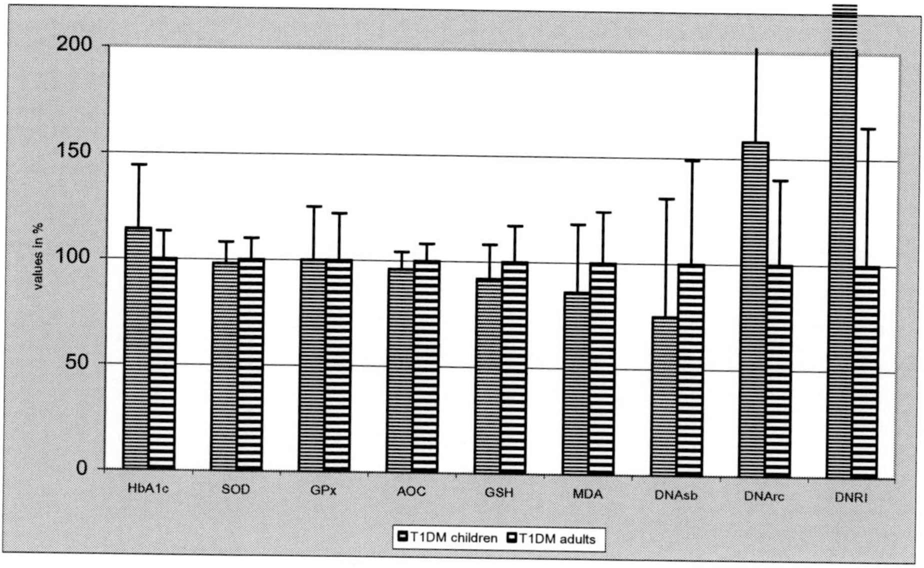

mean values of T1DM adults=100%, DNArc diabetic children SD±49%, DNRI children mean 538%, SD 755%

Figure 30. Markers of oxidative stress, DNA damage and repair.

4.4.4. Comparison of T1DM Adults Divided into 2 Subgroups Based on the Abscence or Presence of Diabetic Microvascular Complications with T1DM Children (Published by Varvarovska, Biomedicineandpharmacotherapy 2004)

In order to support our hypothesis, the group of adults T1DM patients was divided into 2 subgroups. The first one was composed of 23 patients without diabetic microvascular complications, the second group consisted of 30 adults having 1-4 complications (nephropathy, retinopathy, neuropathy, diabetic foot). We compared those 2 subgroups with T1DM children. Though the worse diabetes control (HbA_{1c}) and MDA in children (both values higher on $p<0.05$ versus T1DM adults without DMC), T1DM children displayed similar DNAsb like both adult subgroups. On the contrary, children´s DNArc was significantly higher when compared with T1DM adults without DMC ($p<0.01$) and the comparison with T1DM adults with diabetic complications showed the statistical significance on $p<0.001$. DNRI behaved similarly like DNArc, i.e. statistical significance children versus T1DM without DMC $p<0.01$, and versus T1DM with DMC

$p<0.001$. Hence, it could be concluded that T1DM children had better antioxidant defence against oxidative stress including superior DNA repair process. Adult diabetic patients with microvascular complications showed the worst values of DNA assays and this might agree with our hypothesis, i.e. an individually bad DNA repair could contribute to the earlier development of diabetic microvascular complications. However, it is also necessary to acknowledge a fact, that diabetic patients with DMC had longer disease duration (both subgroups $p<0.001$ versus diabetic children, not shown in the figure), so DNA repair could be exhausted after a long exposition to oxidative stress originated from the long-lasting hyperglycemia.

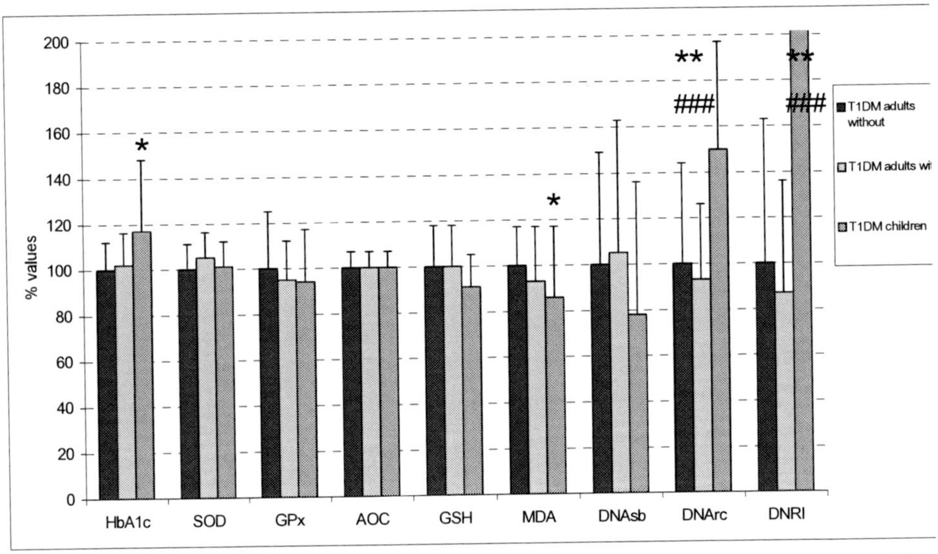

(mean values of adult diabetic patients without microvascular complications=100%) , published by Varvarovska (..)

T1DM adults without= T1DM patients without diabetic microvascular complications (DMC), T1DM with=adults with DMC,

mean values of T1DM adults without complications=100%

*p<0.05 **p<0.01 T1DM children versus adult T1DM patients without diabetic microvascular complications

p<0.001 T1DM children versus adult T1DM with diabetic microvascular complications

DNRI children mean 489%, SD693%

Figure 31. Markers of oxidative stress, DNA damage and repair.

4.4.5. Study of CD Patients (Adults and Children) Versus Healthy Adult Population

The values of SOD, GPx and GSH in CD patients cannot be compared with the findings in GIT mucosa, as markers of oxidative stress were measured in blood or erythrocytes. Nevertheless, it is obvious that CD patients had lower GPx activity, particularly CD children ($p<0.01$) when compared with healthy adults. Also total plasmatic antioxidant capacity was significantly low both in CD adults and children (both $p<0.01$). In CD adults, GSH is significantly higher ($p<0.01$), also their MDA low levels corresponds with the more localized and inflammatory pattern of the disease ($p<0.05$). In children, these two markers are presented in a similar way, but there are not significantly changed.

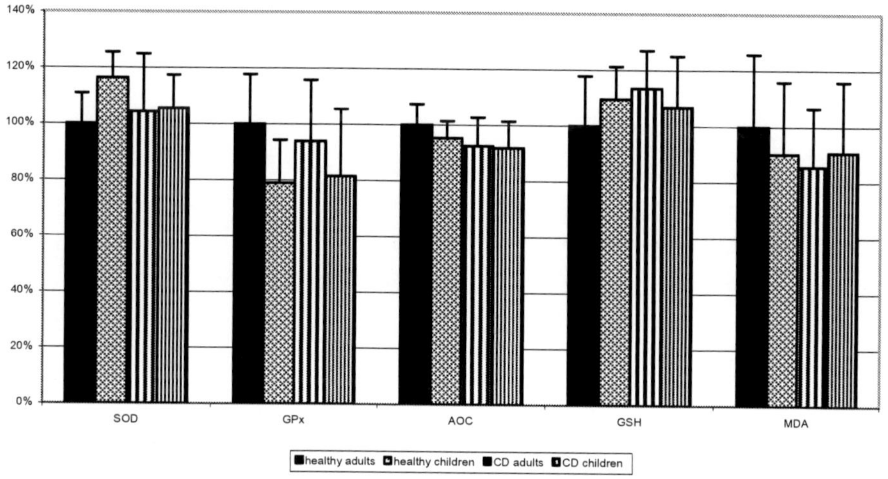

mean values of healthy adults=100%

Figure 32. Markers of oxidative stress.

In correlation with the scarce literary data, our patients had significantly decreased antioxidant plasmatic activity and lower GPx (mainly children) [153]. Lower GPx gives evidence of the worse antioxidant defence of CD patients. The levels of GSH was higher in CD adults (versus healthy adults) what might be in relationship with a special diet applied in those group of patients.

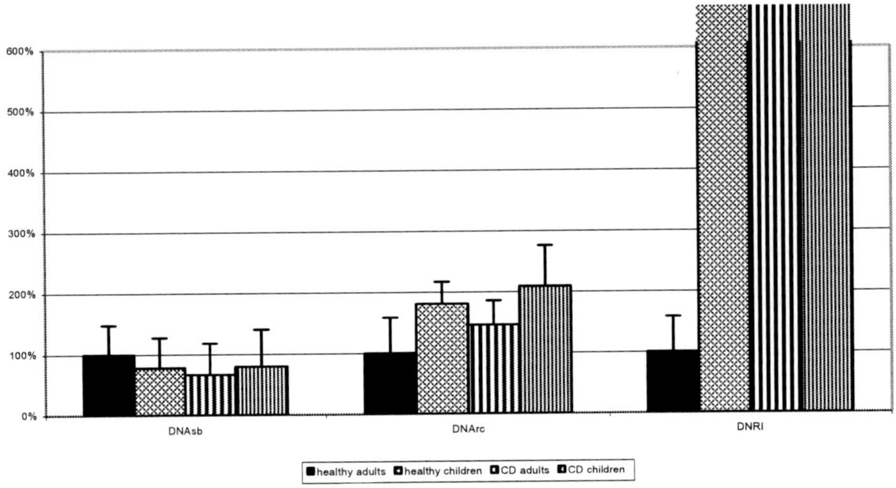

mean values of healthy adults=100%, healthy children DNRI mean 1144%, SD 3026%, CD adults DNRI mean 668%, SD 1256%, CD children DNRI mean 862%, SD 1240%

Figure 33. DNA damage and repair.

CD adults and children showed decreased levels of DNA damage (CD adults DNA sb $p<0.05$ versus healthy adults). On the contrary, the significantly increased DNA repair capacity was found in both groups of CD patients (adults $p<0.05$, children $p<0.001$). Similar finding was confirmed in DNRI (both groups $p<0.001$).

The studied group of CD adults and children did not display increased DNA damage as cited in the literature (157). Their levels of DNA repair capacity was significantly high mainly in children, it corresponds with the quality of young organism and is stimulated in inflammatory disease. These results are contradictory to the findings in idiopathic proctocolitis where DNA repair is usually low (but there is a susceptibility to carcinogenesis)[160, 161, 162, 163]. Our results of DNA repair cannot be compared with similar study in the literature.

4.4.6. Comparison of Adult CD Patients with CD Children

CDAI index was higher in children, but not significantly (large variance). So their disease was more active (we dared to use CDAI index because pediatric patients were at the older school age or adolescents). CD children had shorter disease duration ($p<0.001$, not shown here). Markers of oxidative stress did not differ significantly from those of CD adults. DNAsb and DNRI in children were elevated, but not significantly. The intensity of DNA repair was significantly higher in children than in CD adults ($p<0.05$), certainly reflecting their younger age.

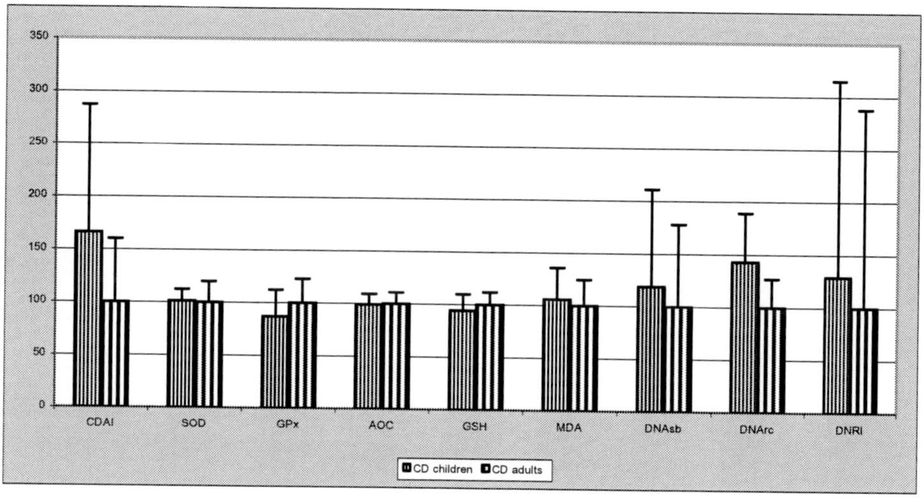

mean values of CD adults=100%

Figure 34. Markers of oxidative stress, DNA damage and repair.

4.4.7. Comparison of Adult Diabetic Patients and Patients with Crohn's Disease

Patients with T1DM were comparable with CD patients with regard to their age, but they had longer disease duration ($p<0.001$). Markers of oxidative stress in T1DM adults displayed lower GSH ($p<0.05$) than CD adults, and also of course very high MDA ($p<0.001$). DNAsb were significantly higher in T1DM adults ($p<0.01$), but their DNArc did not differ substantially. So DNRI in diabetic adults was significantly lower than in CD patients ($p<0.05$).

The presented results displayed the more expressed consequencies of oxidative stress in T1DM patients than in CD adults. It is not excluded that besides the generalized influence of hyperglycemic state, the disease duration (longer in T1DM group) might influence the studied parameters.

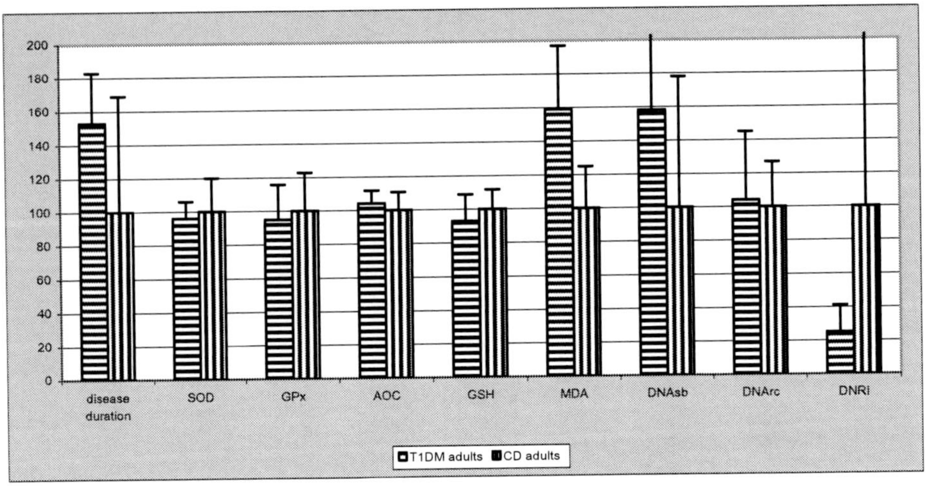

mean values of CD adults=100%, T1DM adults DNAsb SD 78%, CD adults DNRI SD 188%

Figure 35. Markers of oxidative stress, DNA damage and repair.

4.4.8. Children with T1DM and CD Compared with Healthy Children

The children with T1DM had the lowest levels of SOD, stimulated GPx (even if not statistically significant), importantly decreased GSH when compared with healthy children. DNA damage in diabetic children did not differ from the other groups, but DNA repair capacity was statistically increased ($p<0.05$). CD children did not differ from healthy children though similar tendencies in the examined parameters like in diabetic children were observed (SOD, GPX, AOC, GSH, DNArc and DNRI, but not MDA).

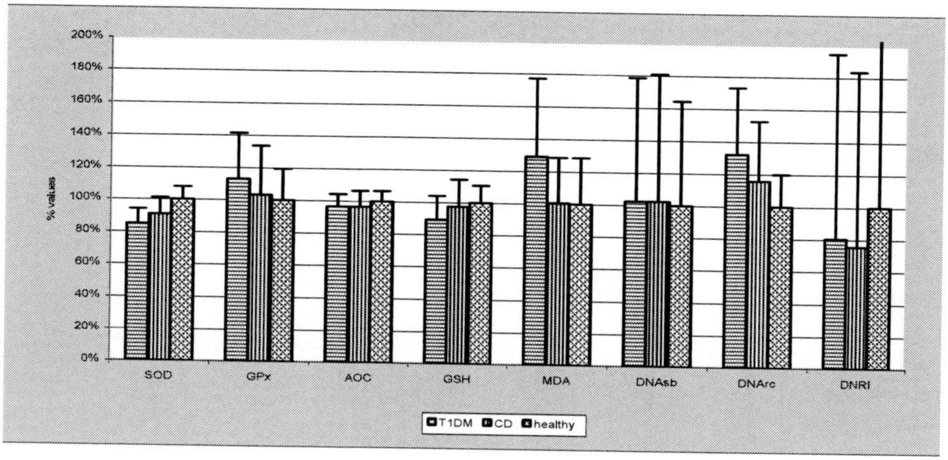

mean values of healthy children=100%, DNRI healthy children SD 264%

Figure 36. Markers of oxidative stress, DNA damage and repair/

Table 1. Comparison of T1DM and CD children versus healthy children

CHILDREN		T1DM	CD	HEALTHY
age	years	13.26±2.98	14.29±2.52	13.73±3.8
disease duration	years	2.94±2.72	2.99±1.43	
HbA1c	% (DCCT - <5%)	9.5±2.52		
PCDAI	Index (<30)		35.69±16.6	
SOD	U/g Hb	1227±131***	1312±148 *	1447±115
GPx	U/g Hb	53.68±13.44	49.05±14.40	47.52±9.21
AOC	mmol/l plasma	1.36±0.11	1.36±0.14	1.41±0.09
GSH	mmol/l ery	2.02±0.34 *	2.21±0.38	2.27±0.24
MDA	μmol/l plasma	2.49±0.93	1.94±0.54	1.93±0.55
DNAsb	breaks per 10^9daltons	0.43±0.32	0.43±0.33	0.42±0.27
DNArc	breaks per 10^9daltons	1.86±0.58 *	1.63±0.52	1.41±0.28
DNRI	index	13.88±19.74	13.02±18.72	17.28±45.69

The results of children with T1DM demostrated the more intensive oxidative stress than in CD and healthy children. It is also obvious that DNA repair was apparently stimulated, and thus correcting the present DNA oxidative damage. Oxidative stress due generalized long-lasting hyperglycemic state overwhelms T1DM children and obviously leads to the more intensive oxidative stress than a patchy granulomatous inflammation in CD children. The results cannot be compared with similar findings in the literature.

4.4.9. Comparison between T1DM and CD Children

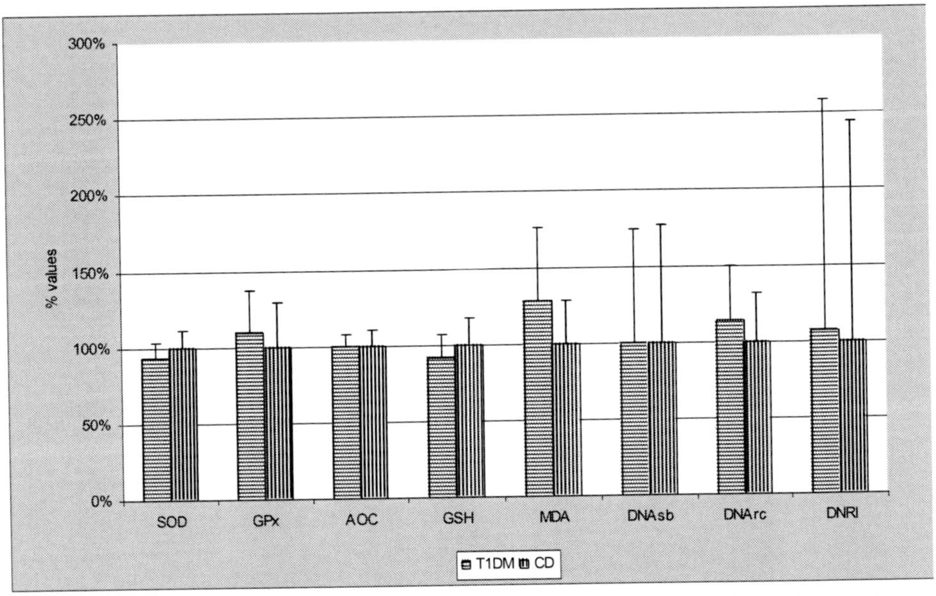

Values of CD children=100%; mean values of CD children=100%; Comparison between T1DM children and CD children did not show any important differences. Both groups were comparable in their age, sex, and disease duration.

Figure 37. Markers of oxidative stress, DNA damage and repair.

4.4.10. Comparison of all Groups of Patients Versus Healthy Adults

The summarizing graph brings a complex view on differences among healthy adults and children, CD adults and children, as well as T1DM adults and children, and finally T2DM adults. Both CD patients had significantly low total antioxidant plasma capacity ($p<0.01$), but on the other hand elevated GSH in CD adults ($p<0.01$). This finding can be influenced by a special diet in the majority of adult CD patients. CD children had SOD elevated only moderately, but significantly lower GPx ($p<0.01$). Similar finding was displayed in diabetic adults and children. In addition, they had also significantly low AOC. MDA was elevated in all T1DM patients in agreement with the disease character. In adult T2DM patients, extremely high SOD was found, and reversely low GSH ($p<0.01$) and MDA ($p<0.01$) as mentioned above.

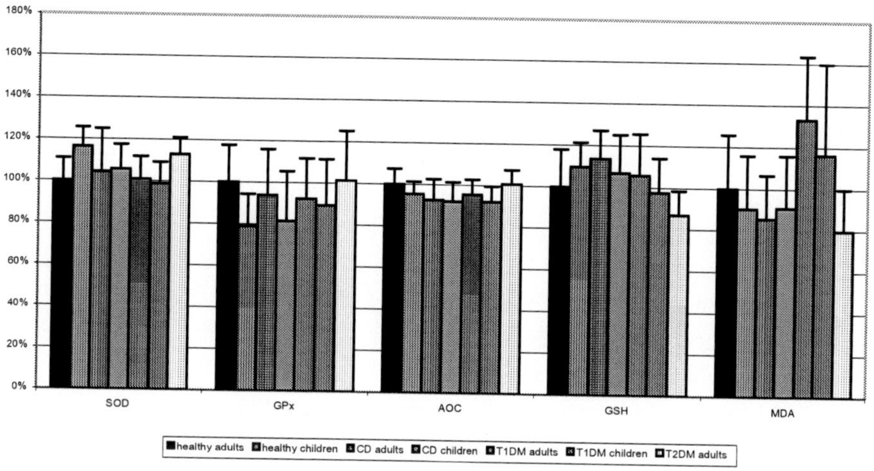

mean values of healthy adults=100%

Figure 38. Markers of oxidative stress.

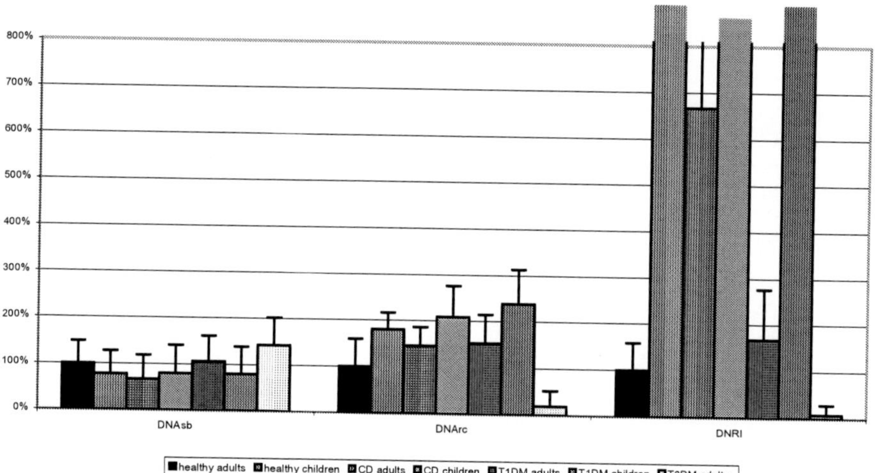

mean values of healthy adults=100%, DNRI healthy children mean 1144%, SD 3026%, DNRI CD adults mean 668%, SD 1256%, DNRI CD mean children 862%, SD 1240%, DNRI T1DM children mean 919%, SD 1307%

Figure 39. DNA damage and repair.

It is evident that all groups of children differed from healthy adults in DNAsb, DNArc and DNRI. They had lower levels of DNAsb, i.e. less DNA damage. On the contrary, DNArc was significantly higher in healthy children ($p<0.001$), CD children ($p<0.001$), and mainly in T1DM children ($p<0.001$). Analogous assay tendency was found both in CD and T1DM adults versus healthy adult controls, i.e. lower DNAsb (CD adults $p<0.05$, similar but not significant finding in T1DM adults), DNArc increased in CD adults ($p<0.05$) and also in T1DM adults ($p<0.001$). DNRI reflected those presented results, i.e. they were elevated both in CD and T1DM adults ($p<0.001$). The completely different trend was observed in T2DM adults that showed high DNAsb ($p<0.05$), extremely low DNA repair capacity ($p<0.001$) and also DNRI ($p<0.001$).

It could be concluded that both children and adults with Crohn´s disease or type 1 diabetes mellitus displayed stimulated DNA repair of oxidative DNA damage. On the other hand, adult patients with T2DM had very low DNA repair capacity and this might be linked with their susceptibility to carcinogenesis.

Chapter V

Conclusion

Oxidative stress as predominance of reactive oxygen species in organism contributes substantially to many pathological processes in human body. An intensive research has brought a developing knowledge of oxidative stress and its impact on human metabolism. The short survey in the first theoretic part of this chapter summarizes current knowledge of oxidative stress, antioxidant defence, new antioxidant therapeutic possibilities. As oxidative stress injures DNA structure, oxidative DNA damage and DNA repair process has been reviewed with regard to different human diseases.

In the second part of this chapter, the authors present their study of oxidative stress, oxidative DNA damage and DNA repair in patients with diabetes mellitus type 1, type 2 and with Crohn´s disease. Summarizing the results, adult patients with type 1 diabetes mellitus had the signs of important oxidative stress, but compensated with a stimulated DNA repair. The subgroup of T1DM with diabetic microvascular complications had lower DNA repair, maybe due to exhaustion of repair machinery in a long disease duration or due to a decreased individual quality of DNA repair. When T1DM adults were compared with T1DM children, the young patients showed better results of parameters of oxidative stress and very high DNA repair, certainly in association with short disease duration and young age, though their diabetes control was not good. The adult patients with type 2 diabetes had very high oxidative DNA damage and extremely low DNA repair, these signs are assumed to be linked to the possible future carcinogenesis. Patients with Crohn´s disease (both children and adults) did not display such an intensive oxidative stress as diabetic patients did. It could be related to the fact that Crohn´s disease need not affect the whole organism like hyperglycemia in diabetes does. The results acquired in T1DM adult patients seems to support our hypothesis i.e. patients with bad DNA repair may have the development of diabetic microvascular complications. Nevertheless, large population studies in DM or IBD studying intensity of OS and expression of DNA repair enzymes are needed in order to get the correlation between individual repair enzymes expression and the long-term course and occurence of complications in DM or IBD.

Acknowledgement

The study was supported by the grant IGA MZ CR NR7919-3/2004 Individual DNA repair capacity induced by oxidative stress and its influence on the development and prognosis of some chronic non-infectious diseases, and the Research Aim MZ CR MSM 0021620814 Prevention, diagnostics, and the therapy of diabetes mellitus, metabolic and endocrine disorders of human organism.

References

[1] Dröge, W. Free Radicals in the Physiological Control of Cell Function. *Physiol Rev,* 2002, 82, 47-95.

[2] Cooke, MS; Evans, MD; Dizdaroglu, M; Lunec, J. Oxidative DNA damage: mechanisms, mutation, and disease. *The FASEB Journal,* 2003, 17, 1195-1214.

[3] Stipek, S. *Antioxidants and free radicals in health and disease.* 1st Edition. Praha, Grada Publishing, 2000.

[4] Racek, J. *Oxidative stress and how to influence it.* 1st Edition. Praha, Galen, 2003.

[5] Forsberg, L; de Faire, U; Morgenstern, R. Oxidative Stress, Human Genetic Variation, and Disease. *Archives of Biochemistry and Biophysics,* 2001, 389, 84-93.

[6] McCord, JM; Edeas, MA. SOD, oxidative stress and human pathologies: a brief history and a future vision. *Biomed Pharmacother,* 2005, 59, 139-42.

[7] Muscoli, C; Cuzzocrea, S; Riley, DP; Zweier, JL; Thiemermann, C; Wang, ZQ; Salvemini, D. On the selectivity of superoxide dismutase mimetics and its importance in pharmacological studies. *British Journal of Pharmacology,* 2003, 140, 445-460.

[8] Day, BJ. Catalytic antioxidants: a radical approach to new therapeutics. *Drug Discovery Today,* 2004, 9, 557-566.

[9] Vouldoukis, I; Conti, M; Krauss, P; Kamaté, C; Blazquez, S; Tefit, M; Mazier, D; Calenda, A; Dugas, B. Supplementation with Gliadin-combined Plant Superoxide Dismutase Extract Promotes Antioxidant Defences and Protects Against Oxidative Stress. *Phytotherapy Research,* 2004, 18, 957-962.

[10] Hayes, JD. Glutathione transferases. *Annu Rev Pharmacol Toxicol*, 2005, 45, 51-88.

[11] Townsend, D; Tew, KD; Tapiero, H. The importance of glutathione in human disease. *Biomed Pharmacother,* 2003, 57, 145-155.

[12] Sgambato, A; Campisi, B; Zupa, A; Bochicchio, A; Romano, G; Tartarone, A; Galasso, R; Traficante, A; Cittadini, A. Glutathione S-transferase (GST) polymorphisms as risk factors for cancer in a highly homogenous population from sosuthern Italy. *Anticancer Research,* 2002, 22, 3647-52.

[13] Dusinska, M; Ficek, A; Horska, A; Raslova, K; Petrovska, H; Vallova, B; Drlickova, M; Wood, SG; Stupakova, A; Gasparovic, J; Bobek, P; Nagyova, A; Kovacikova, Z; Blazicek, P; Liegebel, U; Collins, AR. Glutathione S-transferase polymorphisms influence the level of oxidative DNA damage and antioxidant protection in humans. *Mutat Research,* 2001, 482, 47-55.

[14] Tamer, L; Calkoglu, M; Ates NA; Zildirim, H; Ercan, B; Saritas, E; Ünlü, A; Atik, U. Glutathione-S-transferase gene polymorphisms (GSTT1, GSTM1, GSTP1) as as increased risk factors for asthma. *Respirology,* 2004, 9, 493-498.

[15] Rahman, SH; Nanny, Ch; Ibrahim, K; O´Reilly, D; Larvin, M; Kingsnorth, AJ; McMahon, MJ. Genetic Polymorphisms of GSTT1, GSTM1, GSTP1, MnSOD, and Catalase in Nonhereditary Chronic Pancreatitis: Evidence of Xenobiotic Stress and Impaired Antioxidant capacity. *Digestive Diseases and Sciences,* 2005, 50, 1376-1383.

[16] Rahman, SH; Ibrahim, K; Larvin, M; Kingsnorth, A; McMahon, MJ. Association of antioxidant enzyme gene polymorphisms and glutathione status with severe acute pancreatitis. *Gastroenterology,* 2004, 126, 1312-1322.

[17] Fukagawa, NK; Liang, P; Li, M; Ashikaga, T; Reddz, KR; Krawitt, EL. Glutathione-S-Transferase M1 null genotype in autoimmune hepatitis. *Dig Dis Sci,* 2001, 46, 2080-3.

[18] Yun, BR; El-Sohemy, A; Cornelis, MC; Bae, SC. Glutathione S-transferase M1, T1, and P1 genotypes and rheumatoid arthritis. *J Rheumatol,* 2005, 32, 992-7.

[19] Janssen, YM; Van Houten, B; Borm, PJ; Mossman, BT. Cell and tissue responses to oxidative damage. *Lab Invest,* 1993, 69, 261-274.

[20] Kidd, PM. Glutathione: Systemic Protectant Against Oxidative and Free Radical Damage. *Alternative Medicine Review,* 1997, 2, 155-176.

[21] Nobili, V; Pastore, A; Gaeta, LM; Tozzi, G; Comparcola, D; Sartorelli, MR; Marcellini, M; Bertini, E; Piemonte, F. Glutathione metabolism and antioxidant enzymes in patients affected by nonalcoholic steatohepatitis. *Clinica Chimica Acta,* 2005, 355, 105-111.

[22] Wu, G; Fang, YZ; Yang, S; Lupton, JR; Turner ND. Glutathione Metabolism and Its Implications for Health. *J Nutr,* 2004,134, 489-92.

[23] Valverde, M; Rojas, E; Kala, SV; Kala, G; Lieberman, MW. Survival and cell death in cells constitutively unable to synthesize glutathione. *Mutation Research,* 2005, Epub ahead of print.

[24] Rahman, I; MacNee, W. Oxidative stress and regulation of glutathione in lung inflammation. *Eur Respir J,* 2000, 16, 534-554.

[25] Tiitto, LH; Peltoniemi, MJ; Kaarteenaho-Wiik, RL; Soini, YM; Paakko, PK; Sormunen, RT; Kinnula, VL. Cell-specific regulation of gamma-glutamylcysteine synthetase in human interstitial lung diseases. *Hum Pathol,* 2004, 35, 832-9.

[26] Hudson, VM. New insights into the pathogenesis of cystic fibrosis: pivotal role of glutathione system dysfunction and implications for therapy. *Treat Respir Med,* 2004, 3, 353-63.

[27] Day, BJ; van Heeckeren, AM; Min, E; Velsor, LW. Role for Cystic Fibrosis Transmembrane Conductance Regulator Protein in a Glutathione Response to Bronchopulmonary Pseudomonas Infection. *Infection and Immunity,* 2004, 72, 2045-2051.

[28] Witschi, A; Reddy, S; Stofer, B; Lauterburg, BH. The systemic availability of oral glutathione. *Eur J Clin Pharmacol,* 1992, 43, 667-9.

[29] Shah, P. *Food Sources that Boost Glutathione Naturally.* 2004 . Available from: http://www.1whey2health.com/food-sources-glutathione.htm.

[30] Brok, J; Buckley, N; Gluud, C. Interventions for paracetamol (acetaminophen) overdoses. *Cochrane Database Syst Rev,* 2002, 3, CD003328.

[31] Bagshaw, SM; Ghali, WA. Acetylcysteine for prevention of contrast-induced nephropathy after intravascular angiography: *A systematic review and meta-analysis.* 2004 october. Available from: http://www.biomedcentral.com/1741-7015/2/38

[32] Dekhuijzen, PNR. Antioxidant properties of N-acetylcysteine: their relevance in relation to chronic obstructive pulmonary disease. *Eur Respir J,* 2004, 23, 629-636.

[33] Spapen, HD; Diltoer, MW; Nguyen, DN; Hendrickx, I; Huyghens, LP. Effects of N-acetylcysteine on microalbuminuria and organ failure in acute severe sepsis: results of a pilot study. *Chest,* 2005, 127, 1413-9.

[34] Hein, OV; Őhring, R; Schilling, A; Oellerich, M; Armstrong, VW; Kox, WJ; Spies, C. N-acetylcysteine decreases lactate signal intensities in liver tissue and improves liver function in septic shock patients, as shown by magnetic resonance spectroscopy: extended case report. *Critical Care,* 2004, 8, R66-71.

[35] Del Sorbo, L; Zhang, H. Is there a place for N-acetylcysteine in the treatment of septic shock? *Critical Care,* 2004, 8, 93-5

[36] Grinberg, L; Fibach, E; Amer, J; Atlas, D. N-acetylcysteine amide, a novel cell-permeating thiol, restores cellular glutathione and protects human red blood cells from oxidative stress. *Free Radical Biology and Medicine,* 2005, 38, 136-145.

[37] Das, DK. Thioredoxin regulation of ischemic preconditioning. *Antioxid Redox Signal,* 2004, 6, 405-12.

[38] Nakamura, T; Nakamura, H; Hoshino, T; Ueda, S; Wada, H; Yodoi, J. Redox regulation of lung inflammation by thioredoxin. *Antioxid Redox Signal*, 2005, 7, 60-71.

[39] Alberts, B; Johnson, A; Lewis, J; Raff, M; Roberts, K; Walter, P. *Molecular biology of the cell.* Fourth Edition. New York, Garland Science, TaylorandFrancis Group, 2002.

[40] Okuyama, H; Shimahara, Y; Nakamura, H; Araya, S; Kawada, N; Yamaoka, Y; Yodoi, J. Thioredoxin prevents thioacetamide-induced acute hepatitis. *Comparative Hepatology,* 2004, 3 (Suppl 1), S 6.

[41] Bonakdar, RA; Guarneri, E. Coenzyme Q10. *American Family Physician,* 2005, 72, 1065-70.

[42] Langsjoen, PH. Introduction to coenzyme Q10. 1994. Available from: http://faculty.washington.edu/~ely/coenzq10.html

[43] Haargraves, IP; Duncan, AJ; Heales, SJ; Land, JM. The effect of HMG-CoA reductase inhibitors on coenzyme Q10: possible biochemical/clinical implications. *Drug Saf,* 2005, 28, 659-76.

[44] Wang, XL; Rainwater, DL; Mahaney, MC; Stocker, R. Cosupplementation with vitamin E and coenzyme Q10 reduces circulating markers of inflammation in baboons. *Am J Clin Nutr,* 2004, 80, 649-55.

[45] Carcamo, JM; Pedraza, A; Borquez-Ojeda, O; Zhang, B; Sanchez, R; Golde, DW. Vitamin C Is a Kinase Inhibitor: Dehydroascorbic Acid Inhibits IκBα Kinase β. *Molecular and Cellular Biology,* 2004, 24, 6645-6652.

[46] Hart, JA; Kracoff, G; Ottariono, S; Ullmann-Weil, M. Vitamin E. 2002 april. Available from: http://www.umm.edu/altmed/ConsSupplements/Print/VitaminEcs.html .

[47] Niki, E; Noguchi, N. Dynamics of Antioxidant Action of Vitamin E. *Acc Chem Res,* 2004, 37, 45-51.

[48] Li, HL; Wang, AB; Huang, Y; Liu, DP; Wei, C; Williams, GM; Zhang, CN; Liu, G; Liu, YQ; Hao, DL; Hui, RT; Lin, M; Liang, CC. Isorhapontigenin, a new resveratrol analog, attenuates cardiac hypertrophy via blocking signaling transduction pathways. *Free Radic Biol Med,* 2005, 38, 243-57.

[49] Borek, C. Dietary Antioxidants and Human Cancer. *Integrative cancer therapies,* 2004, 3, 333-341.

[50] Valko, M; Izakovic, M; Mazur, M; Rhodes, CJ; Telser, J. Role of oxygen radicals in DNA damage and cancer incidence. *Molecular and Cellular Biochemistry,* 2004, 266, 37-56.

[51] Delmas, D; Jannin, B; Latruffe, N. Resveratrol: Preventing properties against vascular alterations and ageing. *Mol Nutr Food Res,* 2005, 49, 377-395.

[52] Wang, QL; Lin, M; Liu, GT. Antioxidative activity of natural isorhapontigenin. *Jpn J Pharmacol,* 2001, 87, 61-66.

[53] Bernstein, C; Bernstein, H; Payne, CM; Garewal, H. DNA repair/pro-apoptotic dual-role proteins in five major DNA repair pathways: fail-safe protection against carcinogenesis. *Mutation Research,* 2002, 511, 145-178.

[54] Hansen, WK; Kellez, MR. Review of Mammalian DNA Repair and Translational Implications. *The Journal of Pharmacology and Experimental Therapeutics,* 2000, 295, 1-9.

[55] Smith, PJ; Jones, CJ. *DNA Recombination and Repair.* 1st Edition, New York, Oxford University Press, 1999.

[56] Wood, RD; Mitchell, M; Sgouros, J; Lindahl, T. Human DNA Repair Genes. *Science,* 2001, 291, 1284-1289.

[57] Wood, RD; Mitchell, M; Lindahl, T. Human DNA repair genes. *Mutat Res,* 2005, 577, 275-83.

[58] Huberman JA. *DNA repair pathways. Lectures.* 2005, October-November. Available from: http://saturn.roswellpark.org/cmb/huberman/DNA-Repair/lecture-intro.html.

[59] Herbert, KE; Mistry P. DNA Oxidation: Basics of Repair. *Virtual Free Radical School.* 2003, May. Available from: www.medicine.uiowa.edu/frrb/V irtualSchool/Herbert-DNA-Repair.ppt.

[60] Passarge E. *Color Atlas of Genetics.* 2nd Edition. New York., Thieme New York, 2001.

[61] Thompson, LH; Schild, D. Recombinational DNA repair and human disease. *Mutation Research,* 2002, 509, 49-78.

[62] Duker, NJ. Chromosome breakage syndromes and cancer. *Am J Med Genet,* 2002, 115, 125-9.

[63] Welcsh, PL; King, MC. BRCA1 and BRCA2 and the genetics of breast and ovarian cancer. *Human Molecular Genetics,* 2001, 10, 705-713.

[64] Evans, MD; Dizdaroglu, M; Cooke, MS. Oxidative DNA damage and disease: induction, repair and significance. *Mutation Research,* 2004, 567, 1-61.

[65] Collins, A; Dusinska, M; Franklin, M; Somorovska, M; Petrovska, H; Duthie, S; Fillion, L; Panayiotidis, M; Raslova, K; Vaughan, N. Comet assay in human biomonitoring studies: reliability, validation, and applications. *Environ Mol Mutagen,* 1997, 30, 139-46.

[66] Powell, CL; Swenberg, JA; Rusyn, I. Expression of base excision DNA repair genes as a biomarker of oxidative DNA damage. *Cancer Letters,* 2005, 229, 1-11.

[67] Cheadle, JP; Sampson, JR. Exposing the MYtH about base excision repair and human inherited disease. *Human Molecular Genetics,* 2003, 12, R159-R165.

[68] Fukae, J; Takanashi, M; Kubo, S; Nishioka, K; Nakabeppu, Y; Mori, H; Mizuno, Y; Hattori, N. Expression of 8-oxoguanine DNA glycosylase (OGG1) in Parkinson´s disease and related neurodegenerative disorders. *Acta Neuropathologica,* 2005, 109, 256-62.

[69] Mathonnet, G; Labuda, D; Meloche, C; Wambach, T; Krajinovic, M; Sinnet, D. Variable continental distribution of polzmorphisms in the coding regions of DNA-repair genes. *J Hum Genet,* 2003, 48, 659-664.

[70] Collins, AR; Dusinska, M; Horvathova, E; Munro, EW; Savio, M; Stetina, R. Inter-individual differences in repair of DNA base oxidation, measured in vitro with the comet assay. *Mutagenesis,* 2001, 16, 297-301.

[71] Stevnsner, T; Thorslund, T; de Souza/Pinto, NC; Bohr, VA. Mitochondrial repair of 8-oxoguanine and changes with aging. *Experimental Gerontology,* 2002, 37, 1189-1196.

[72] Larsen, NB; Rasmussen, M; Rasmussen, LJ. Nuclear and mitochondrial DNA repair: similar pathways? *Mitochondrion,* 2005, 5, 89-108.

[73] Bürkle, A. Physiology and pathophysiology of poly(ADP-ribosyl)ation. *BioEssays,* 2001, 23, 795-806.

[74] Virag, L. Structure and Function of Poly(ADP-ribose) Polymerase-1: Role in Oxidative Stress-Related Pathologies. *Current Vascular Pharmacology,* 2005, 3, 209-214.

[75] Erdelyi, K; Bakondi, E; Gergely, P; Szabo, C; Virag, L. Pathophysiologic role of oxidative stress-induced poly(ADP-ribose) polymerase-1 activation: focus on cell death and transcriptional regulation. *Cell Mol Life Sci,* 2005, 62, 751-759.

[76] Wiesmüller, L. Genetic stabilization by p53 involves growth regulatory and repair pathways. *Journal of Biomedicine and Biotechnology,* 2001, 1, 7-10.

[77] Beneke, S; Diefenbach, J; Bürkle, A. Poly(ADP-ribosyl)ation inhibitors: promising drug candidates for a wide variety of pathophysiologic conditions. *Int J Cancer,* 2004, 111, 813-818.

[78] Woon, EC; Threadgill, MD. Poly(ADP-ribose)polymerase inhibition – where now? *Curr Med Chem,* 2005, 12, 2373-92.

[79] Szabo, C. Pharmacological inhibiton of poly(ADP-ribose)polymerase in cardiovascular disorders: future directions. *Curr Vasc Pharmacol,* 2005, 3, 301-303.

[80] Inotek News. 2003, April. Available from: www.inotekcorp.com/news/apr-26-2003.htm

[81] De Magalhaes, JP. From cells to ageing: a review of models and mechanisms of cellular senescence and their impact on human ageing. *Experimental Cell Research,* 2004, 300, 1-10.

[82] Chen, SK; Hsieh, WA; Tsai, MH; Chen, CH; Hong, AI; Wei, YH; Chang, WP. Age-associated Decrease of Oxidative Repair Enzymes, Human 8-Oxoguanine DNA Glycosylases (hOgg1), in Human Aging. *J Radit Res,* 2003, 44, 31-35.

[83] Szczesny, B; Bhakat, KK; Mitra, S; Boldogh, I. Age-dependent modulation of DNA repair enzymes by covalent modification and subcellular distribution. *Mechanisms of Ageing and Development,* 2004, 125, 755-765.

[84] Szczesny, B; Hazra, TK; Papaconstantinou, J; Mitra, S; Boldogh, I. Age-dependent deficiency in import of mitochondrial DNA glycosylases required for repair of oxidatively damaged bases. *PNAS,* 2003, 100, 10670-10675.

[85] Lee, JW; Harrigan, J; Opresko, PL; Bohr, VA. Pathways and functions of the Werner syndrome protein. *Mechanisms of Ageing and Development,* 2005, 126, 79-86.

[86] Mohaghegh, P; Hickson, ID. Premature aging in RecQ helicase-deficient human syndromes. *The International Journal of BiochemistryandCell Biology,* 2002, 34, 1496-1501.

[87] Opresko, PL; Cheng, WH; von Kobbe, C; Harrigan, JA; Bohr, VA. Werner syndrome and the function of the Werner protein; what they can teach us about the molecular aging process. *Carcinogenesis,* 2003, 24, 791-802.

[88] Browner, WS; Kahn, AJ; Ziv, E; Reiner, AP; Oshima J; Cawthon, RM; Hsueh, WC; Cummings, SR. The genetics of Human Longevity. *Am J Med,* 2004, 117, 851-860.

[89] Szibor, M; Holtz, J. Mitochondrial ageing. *Basic Res Cardiol,* 2003, 98, 210-218.

[90] Wei, YH; Lee, HC. Oxidative stress, Mitochondrial DNA Mutation, and Impairment of Antioxidant Enzymes in Aging. *Exp Biol Med,* 2002, 227, 671-682.

[91] Sastre, J; Pallardo, FV; Vina, J. The Role of Mitochondrial Oxidative Stress in Aging. *Free Radical BiologyandMedicine,* 2003, 35, 1-8.

[92] Caporaso N. The Molecular Epidemiology of Oxidative Damage to DNA and Cancer. *J Natl Cancer Inst,* 2003, 95, 1263-5.

[93] Dixon, K; Kopras, E. Genetic alterations and DNA repair in human carcinogenesis. *Seminars in Cancer Biology,* 2004, 14, 441-448.

[94] Collins, A; Harringtion, V. Repair of oxidative DNA damage: assessing its contribution to cancer prevention. *Mutagenesis,* 2002, 17, 489-493.

[95] Shackelford, ER; Manuszak, RP; Heard, SC; Link, CJ; Wang, S. Pharmacological manipulation of ataxia-teleangiectasia kinase activity as a treatment for Parkinson´s disease. *Med Hypotheses,* 2005, 64, 736-41.

[96] Martinet, W; Knaapen, MWM; De Meyer, GRY; Herman, AG; Kockx, MM. Elevated Levels of Oxidative DNA Damage and DNA Repair Enzymes in Human Atherosclerotic Plaques. *Ciruclation,* 2002, 106, 927-932.

[97] Martinet, W; Knaapen, MWM; De Meyer, GRY; Herman, AG; Kockx, MM. Oxidative DNA Damage and Repair in Experimental Atherosclerosis Are Reversed by Dietary Lipid Lowering. *Circ Res,* 2001, 88, 733-739.

[98] Hofseth, LJ; Khan, MA; Ambrose, M; Nikolayeva, O; Xu-Welliver, M; Kartalou, M; Hussain, SP; Roth, RB; Zhou, X; Mechanic, LE; Zurer, I; Rotter, V; Samson, LD; Harris, CC. The adaptive imbalance in base excision-repair enzymes generates microsatellite instability in chronic inflammation. *J Clin Invest,* 2003, 112, 1887-94. Erratum in: J Clin Invest, 2004, 113, 490.

[99] Blasiak, J; Arabski, M; Krupa, R; Wozniak, K; Zadrozny, M; Kasznicki, J; Zurawska, M; Drzewoski, J. DNA damage and repair in type 2 diabetes mellitus. *Mutat Res,* 2004, 554, 297-304.

[100] Miyako, K; Kohno, H; Ihara, K; Kuromaru, R; Matsuura, N; Hara, T. Association study of human MTH1 gene polymorphisms with type 1 diabetes mellitus. *Endocr J,* 2004, 51, 493-8.

[101] Vodicka, P; Koskinen, M; Stetina, R; Soucek, P; Vodickova, L; Matousu, Z; Kuricova, M; Hemminki, K. The role of various biomarkers in the evaluation of styrene genotoxicity. *Cancer Detect Prev,* 2003, 27, 275-84.

[102] The Expert Committee on the Diagnosis and Classification of Diabetes Mellitus. Report of the Expert Committee on the Diagnosis and Classification of Diabetes Mellitus. *Diabetes Care,* 2003, 26, S 5-20.

[103] Brownlee, M. Glycation products and the pathogenesis of diabetic complications. *Diabetes Care,* 1992, 15, 1835-1843.

[104] Brownlee, M. Biochemistry and molecular cell biology of diabetic complications. *Nature,* 2001, 414, 813-820.

[105] Brownlee, M. The Pathobiology of Diabetic Complications. *Diabetes,* 2005, 54, 1615-1625.

[106] Ceriello, A. New insights on oxidative stress and diabetic complications may lead to a "causal" antioxidant therapy. *Diabetes Care,* 2003, 26, 1589-96.

[107] Skyler, JS. *Atlas of Diabetes.* 2nd Edition. Philadelphia, Lippincott, WilliamsandWilkins, 2002.

[108] VanderJagt, DJ; Harrison, JM; Ratliff, DM; Hunsaker, LA; Vander Jagt, DL. Oxidative stress indices in IDDM subjects with and without long-term diabetic complications. *Clinical Biochemistry,* 2001, 34, 265-270.

[109] Holeček, V; Racek, J. The importance of free radicals and advanced glycosylation end-products (AGE) in diabetes mellitus. *Klin Biochem Metab,* 1995, 3, 18-22.

[110] Jakus, V. The role of free radicals, oxidative stress and antioxidant systems in diabetic vascular disease. *Bratisl Lek Listy* 2000, 101, 541-551.

[111] Jakus, V; Carsky, J; Hrnciarova, M. Importance of advanced glycosylation end-products AGE. *Bratisl Lek Listy,* 1998, 99, 368-375.

[112] Jakus, V; Rietbrock, N. Advanced Glycation End-Products and the Progress of Diabetic Vascular Complications. *Physiol. Res,* 2004, 53, 131-142.

[113] Vlassaara, H. Recent progression in the biologic and clinical significance of advanced glycosylation end products. *J Lab Clin Med* 1994, 124, 19-30.

[114] Rahbar, S; Figarola, JL. Inhibitors and Breakers of Advanced Glycation Endproducts (AGEs): A Review. *Curr Med Chem-Immun, EndocandAgents,* 2002, 2, 135-161.

[115] Kalousova, M; Zima, T; Tesar, V; Dusilova/Sulkova, S; Skrha, J. Advanced glycoxidation end products in chronic disease – clinical chemistry and genetic background. *Mutation Research,* 2005, 579, 37-46.

[116] Jenkins, AJ; Best, JD; Klein, RL; Lyons, TJ. Lipoproteins, glycoxidation and diabetic angiopathy. *Diabetes Metab Res Rev,* 2004, 20, 349-368.

[117] Aljada, A; Ghanim, H; Mohanty, P; Syed, T; Bandyopadhyay, A; Dandona, P. Glucose intake induces an increse in activator protein 1 and early growth response 1 binding activities, in teh expression of tissue factor and matrix metalloproteinase in mononuclear cells, and in plasma tissue factor and matrix metalloproteinase concentrations. *Am J Clin Nutr,* 2004, 80, 51-7.

[118] Tsukuhara, H; Sekine, K; Uchiyama, M; Kawakami, H; Hata, I; Todoroki, Y; Hiraoka, M; Kaji, M; Yorifuji, T; Momoi, T; Yoshihara, K; Beppu, M; Mayumi, M. Formation of Advanced Glycosylation End Products and Oxidative Stress in Young Patients with Type 1 Diabetes. *Pediatric Research,* 2003, 54, 419-424.

[119] Hempe, JM; Gomez, R; McCarter, RJ Jr; Chalew, SA. High and low hemoglobin glycation phenotypes in type 1 diabetes. a challenge for interpretation of glycemic control. *J Diabetes Complications,* 2002, 16, 313-320.

[120] Bartos, V; Pelikanova, T. et al. *Diabetology in practice.* 3rd Edition, Prague, Maxdorf-Jessenius, 2003.

[121] Svacina, S. Vascular complications of diabetes. *Trends in current diabetology.* 1st Edition, Prague, Galen, 2005.

[122] Williams, ME. New therapies for advanced glycation end product nephrotoxicity: Current challenges. *American Journal of Kidney Diseases,* 2003, 41 (S1), S42-47.

[123] Chistyakov, DA; Savostanov, KV; Zotova, EV; Nosikov, VV. Polymorphisms in the Mn-SOD and EC-SOD genes and their relationship to diabetic neuropathy in type I diabetes mellitus. *BMC Medical Genetics,* 2001, 2, 4.

[124] Ceriello, A. M; Morocutti, A; Mercuri, F; Quagliaro, L; Moro, M; Damante, G; Viberti GC. Defective Intracellular Antioxidant Enzyme Production in Type 1 Diabetic Patients With Nephropathy. *Diabetes,* 2000, 49, 2170-2177.

[125] Varvarovska, J; Racek, J; Stozicky, F; Soucek, J; Trefil, L; Pomahacova, R. Parameters of oxidative stress in children with Type 1 diabetes mellitus and their relatives. *J Diabetes Complications,* 2003, 17, 7-10.

[126] Krapfenbauer, K; Bimbacher, R; Vierhapper, H; Herkner, K; Kampel, D; Lubec, G. Glycoxidation, and protein and DNA oxidation in patients with diabetes mellitus. *Clinical Science,* 1998, 95, 331-7.

[127] Collins, AR; Raslova, K; Somorovska, M; Petrovska, H; Ondrusova, A; Vohnout, B; Fabry, R; Dusinska, M. DNA damage in diabetes: correlation with a clinical marker. *Free Radic Biol Med,* 1998, 25, 373-7.

[128] Wu, LL; Chiou, CC; Chang, PY; Wu, JT. Urinary 8-OHdG: a marker of oxidative stress to DNA and a risk factor for cancer, atherosclerosis and diabetics. *Clin Chim Acta,* 2004, 339, 1-9.

[129] Samcova, E; Kvasnicova, V. Oxidative stress and 8-hydroxy-2′-deoxyguanosine as a biomarker of oxidative DNA damage. *DMEV,* 2000, 3, 194-198.

[130] Dincer, Y; Akcay, T; Ilkova, H; Alademir, Z; Ozbay, G. DNA damage and antioxidant defense in peripheral leukocytes of patients with Type I diabetes mellitus. *Mutat Res,* 2003, 527, 49-55.

[131] Sardas, S; Yilmaz, M; Oztok, U; Cakir, N; Karakaya, AE. Assessment of DNA strand breakage by comet assay in diabetic patients and the role of antioxidant supplementation. *Mutat Res,* 2001, 490, 123-129.

[132] Du, X; Matsumura, T; Edelstein, D; Rossetti, L; Zsengeller, Y; Szabo, C; Brownlee, M. Inhibition of GAPDH activity by poly(ADP-ribose)polymerase activates three major pathways of hyperglycemic damage in endothelial cells. *J Clin Invest,* 2003, 112, 1049-1057.

[133] Zheng, L; Szabo, C; Kern, TS. Poly(ADP-Ribose) Polymerase Is Involved in the Development of Diabetic Retinopathy via Regulation of Nuclear Factor-κB. *Diabetes,* 2004, 53, 2960-2967.
[134] Miyako, K; Kohno, H; Ihara, K; Kuromaru, R; Matsuura, N; Hara, T. Association Study of Human *MTH1* Gene Polymorphisms with Type 1 Diabetes mellitus. *Endocrine Journal,* 2004, 51, 493-498.
[135] Tyrberg, B; Anachkov, KA; Dib, SA; Wang-Rodriguez, J; Yoon, KH; Levine, F. Islet expression of the DNA repair enzyme 8-oxoguanosine DNA glycosylase (OggI) in human type 2 diabetes. *BMC Endocrine Disorders,* 2002, 2, 2.
[136] Varvarovska, J; Racek, J; Stetina, R; Sykora, J; Pomahacova, R; Rusavy, Z; Lacigova, S; Trefil, L; Siala, K; Stozicky, F. Aspects of oxidative stress in children with type 1 diabetes mellitus. *BiomedandPharmacother,* 2004, 58, 539-45.
[137] Schultz Johansen, J; Harris, AK; Rychly, DJ; Ergul, A. Oxidative stress and the use of antioxidants in diabetes: Linking basic science to clinical practice. *Cardiovascular Diabetology,* 2005, 4, 5.
[138] Da Ros, R; Asaloni, R; Ceriello, A. Antioxidant therapy in diabetic complications: what is new? *Curr Vasc Pharmacol,* 2004, 2, 335-41.
[139] Da Ros, R; Asaloni, R; Ceriello, A. Molecular targets of diabetic vascular complications and potential new drugs. *Curr Drug Targets,* 2005, 6, 503-9.
[140] Reusch, JEB. Diabetes, microvascular complications, and cardiovascular complications: what is it about glucose? *J Clin Invest,* 2003, 112, 986-988.
[141] Li, F; Drel, VR; Szabo, C; Stevens, MJ; Obrosova, IG. Low-Dose Poly(ADP-Ribose) Polymerase Inhibitor-Containing Combination Therapies Reverse Early Peripheral Diabetic Neuropathy. *Diabetes,* 2005, 54, 1514-1522.
[142] Obrosova, IG; Pacher, P; Szabo, C; Zsengeller, Z; Hirooka, H; Stevens, MJ; Yorek, MA. Aldose Reductase Inhibition Counteracts Oxidative-Nitrosative Stress and Poly(ADP-Ribose) Polymerase Activation in Tissue Sites for Diabetes Complications. *Diabetes,* 2005, 54, 234-242.
[143] Shanahan, F. Crohn´s disease. *The Lancet,* 2002, 359, 62-69.
[144] Matulka, M. Immunology of inflammatory bowel diseases. *Alergie,* 2001, 3, 65-70.
[145] Ruf, G. Surgical therapy of complicated inflammatory bowel disease. *Schweiz Rundsch Med Prax,* 2005, 94, 1681-5.

[146] Kruidenier, L; Verspaget HW . Oxidative stress as a pathogenic factor in inflammatory bowel disease—Radicals or ridiculous? *Aliment Pharmacol Ther,* 2002, 16, 1997– 2015.

[147] Kruidenier, L; Kuiper, I; Lamers, CB; Verspaget, HW. Intestinal oxidative damage in inflammatory bowel disease: semi-quantification, localization, and association with mucosal antioxidants. *J Pathol,* 2003, 201, 28-36.

[148] Tuzun, A; Erdil, A; Inal, V; Aydin, A; Bagci, S; Yesilova, Y; Sayal, A; Karaeren, N; Dagalp, K. Oxidative stress and antioxidant capacity in patients with inflammatory bowel disease. *Clin Biochem,* 2002, 35, 569-72.

[149] Yamada, T; Grisham, MB. Role of neutrophil-derived oxidants in the pathogenesis of intestinal inflammation. *Klin Wochenschr,* 1991, 69, 988-94.

[150] Kruidenier, L; Kuper, I; van Duijn, W; Marklund, SL; van Hogezand, RA; Lamers, CB; Verspaget, HW. Differential mucosal expression of three superoxide dismutase isoforms in inflammatory bowel disease. *J Pathol,* 2003, 201, 7-16.

[151] Kruidenier, L; Kuiper, I; Van Duijn, W; Mieremet-Oooms, MA; van Hogezand, RA; Lamers, CB; Verspaget, HW. Imbalanced secondary mucosal antioxidant response in inflammatory bowel disease. *J Pathol,* 2003, 201, 17-27

[152] Wendland, BE; Aghdassi, E; Tam, C; Carrrier, J; Steinhart, AH; Wolman, SL; Baron, D; Allard, JP. Lipid peroxidation and plasma antioxidant micronutrients in Crohn disease. *Am J Clin Nutr,* 2001, 74, 259-64.

[153] Koutroubakis, IE; Malliaraki, N; Dimoulios, PD; Karmiris, K; Castanas, E; Kouroumalis, EA. Decreased total and corrected antioxidant capacity in patients with inflammatory bowel disease. *Dig Dis Sci,* 2004, 49, 1433-7.

[154] Liu, C; Russell, RM; Smith, DE; Bronson, RT; Milburz, PE; Furukawa, S; Wang, XD; Blumberg, JB. The effect of dietary glutathione and coenzyme Q10 on the prevention and treatment of inflammatory bowel disease in mice. *Int J Vitam Nutr Res,* 2004, 74, 74-85.

[155] Oz, HS; Chen, TS; McCLain, CJ; de Villiers, WJ. Antioxidants as novel therapy in a murine model of colitis. *J Nutr Biochem,* 2005, 16, 297-304.

[156] Akgun, E; Caliskan, C; Celik, HA; Ozutemiz, AO; Tuncyurek, M; Aydin, HH. Effects of N-acetylcysteine Treatment on Oxidative Stress in Acetic Acid-induced Experimental Colitis in Rats. *The Journal of International Medical Research,* 2005, 33, 196-206.

[157] D´Odorico, A; Bortolan, S; Cardin, R; D´Inca´, R; Martines, D; Ferronato, A; Sturniolo, GC. Reduced plasma antioxidant concentrations and increased oxidative DNA damage in inflammatory bowel disease. *Scan J Gastroenterol,* 2001, 36, 1289-94.

[158] Seril, DN; Liao, J; Yang, GY; Yang, CS. Oxidative stress and ulcerative colitis-associated carcinogenesis: studies in humans and animal models. *Carcinogenesis,* 2003, 24, 353-362.

[159] D´Inca, R; Cardin, R; Benazzato, L; Angriman, I; Martines, D; Sturniolo, GC. Oxidative DNA damage in the mucosa of ulcerative colitis increases with disease duration and dysplasia. *Inflamm Bowel Dis,* 2004, 10, 23-7.

[160] Sanford, KK; Price, FM; Brodeur, C; Makrauer, FL; Parshad, R. Deficient DNA repair in chronic ulcerative colitis. *Cancer Detect Prev,* 1997, 21, 540-5.

[161] Pokorny, RM; Hofmeister, A; Galandiuk, S; Diety, AB; Cohen, ND; Neibergs, HL. Crohn´ s disease and ulcerative colitis are associated with the DNA repair gene MLH1. *Ann Surg,* 1997, 225, 718-23.

[162] Hofseth, LJ; Khan, MA; Ambrose, M; Nikolayeva, O; Xu-Welliver, M; Kartalou, M; Hussain, SP; Roth, RB; Zhou, X; Mechanic, LE; Zurer, I; Rotter, V; Samson, LD; Harris CC. The adaptive imbalance in base excision-repair enzymes generates microsatellite instability in chronic inflammation. J Clin Invest, 2003, 112, 1887-94. Erratum in: *J Clin Invest,* 2004, 113, 490.

[163] Guo, HH; Loeb, LA. Tumbling down a different pathway to genetic instability. *J Clin Invest,* 2003, 112, 1793-5.

[164] Krishna, M; Woda, B; Savas, L; Baker, S; Banner, B . Expression of p53 antigen in inflamed and regenerated mucosa in ulcerative colitis and Crohn´ s disease. *Mod Pathol,* 1995, 8, 654-7.

[165] Collins AR, Dusinska M, Horvathova E, Munro EW, Savio M, Stětina R. Inter-individual differences in repair of DNA base oxidation, measured in vitro with the comet assay. *Mutagenesis,* 2001, 16, 297-301.

[166] Racek, J; Eiselt, J; Opatrny, K Jr. Transitional elements and free radicals. *Cas Lek Cesk,* 2002 141, 479-82.

[167] Comet Assay Interest Group. Comet assay (single cell gel electrophoresis SCGE). 2005. Available from: http://www.comet.itrcindia.org

[168] Rovnakova, D. *Oxidative stress in children after physical activity – thesis.* Faculty of Medicine, UPJS, Kosice, Slovakia, 2002.

[169] Aguilo, A; Tauler, P; Fuentespina, E; Tur, JA; Cordova, A; Pons, A. Antioxidant response to oxidative stress induced by exhaustive exercise. *PhysiologyandBehavior,* 2005, 84, 1-7.

[170] Barja, G. Free radicals and aging. *TRENDS in Neurosciences,* 2004, 27, 595-600.

[171] Dufour, E; Larsson, NG. Understanding aging: revealing order out of chaos. *Biochim Biophys Acta,* 2004, 1658, 122-32.

[172] Weirich-Schwaiger, H; Weirich, HG; Gruber, B; Schweiger, M; Hirsch-Kauffmann, M. Correlation between senescence and DNA repair in cells from young and old individuals and in premature aging syndromes. *Mutat Res,* 1994, 316, 37-48.

[173] Raji, NS: Surekha, A; Rao, KS. Improved DNA-repair parameters in PHA –stimulated peripheral blood lymphocytes of human subjects with low body mass index. *Mech Ageing Dev,* 1998, 104, 133-48.

[174] Chen, SK; Hsieh, WA; Tsai, MH; Chen, ChCh; Hong, AI; Wei, YH; Chang, WP. Age-associated Decrease of Oxidative Repair Enzymes, Human 8-Oxoguanine DNA Glycosylases (hOgg1) in Human Aging. *J Radiat Res,* 2003, 44, 31-35.

[175] Dincer Y, Akcay T, Alademir Z, Ilkova H.: Effect of oxidative stress on glutathione pathway in red blood cells from patients with insulin-dependent diabetes mellitus. *Metabolism,* 2002, 51, 1360-1362.

[176] Ruiz C, Alegria A, Barbera R, Farre R, Lagard MJ: Lipid peroxidation and antioxidant enzyme activities in patients with type 1 diabetes mellitus. *Scand J Clin Lab Invest,* 1999, 59, 99-105.

[177] Haberland, ME; Fogelman, AM; Edwards, PA. Specifity of receptor-mediated recognition of malondialdehyde-modified low density lipoproteins. *Proc Natl Acad Sci,* 1982, 79, 1712-1716.

Index

B

D

E

F

N

O

P

Q

R

S

T

U

V

W

X

Y

Z

B